3rd International Conference on Materials Engineering and Nano Sciences (ICMENS 2019)

IOP Conference Series: Materials Science and Engineering
Volume 625

Hiroshima, Japan
26 – 28 March 2019

ISBN: 978-1-5108-9829-5
ISSN: 1757-8981

Printed from e-media with permission by:

Curran Associates, Inc.
57 Morehouse Lane
Red Hook, NY 12571

Some format issues inherent in the e-media version may also appear in this print version.

Institute of Physics
Dirac House, Temple Back
Bristol BS1 6BE UK

Phone: 44 1 17 929 7481
Fax: 44 1 17 920 0979

techtracking@iop.org

Additional copies of this publication are available from:

Curran Associates, Inc.
57 Morehouse Lane
Red Hook, NY 12571 USA
Phone: 845-758-0400
Fax: 845-758-2633
Email: curran@proceedings.com
Web: www.proceedings.com

Table of contents

Volume 625

2019 the 3rd International Conference on Materials Engineering and Nano Sciences (ICMENS 2019)

26–28 March 2019, Hiroshima, Japan

Accepted papers received: 27 August 2019
Published online: 30 September 2019

Preface

Papers

Material structure and properties

Material applications

IOP Conf. Series: Materials Science and Engineering **625** (2019) 011001 doi:10.1088/1757-899X/625/1/011001

Preface

Materials Engineering based on Nano Sciences is a key ingredient of current advanced technologies supporting modern society, especially Information and Energy Technologies. The tremendous rapid growth of these fields makes them highly interdisciplinary and then inevitably demands their mutual interactions. Under these circumstances, the 2019 the 3rd International Conference on Materials Engineering and Nano Sciences (ICMENS 2019) was organized as a series of successful international conferences held in Singapore (2017) and Hong Kong (2018).

The main goal of ICMENS 2019, which will be held from 26 to 28 March 2019, in Hiroshima, is to address latest original results in Materials Engineering and Nano Sciences, including both theoretical advances and practical implementations, which are becoming more and more popular in industry and in our daily lives. The ICMENS 2019 will provide a premier interdisciplinary platform for scientists, researchers, industry leaders, engineers and educators throughout the world to present and discuss the most recent innovations, trends, concerns, as well as practical challenges encountered, and streamline solutions in the fields of Materials Engineering and Nano Sciences. The meeting will provide an opportunity to highlight recent developments and to identify emerging and future areas of growth in Quantum Technology.

Towards this end, the Technical Programme Committee has assembled an excellent programme comprising of 2 excellent Keynote Speeches and 2 Plenary Talks from renowned scientists from the world, 6 parallel technical sessions comprising of more than 40 presentations. A total of 31 full papers were selected from 46 original contributions after a rigorous review process. In addition to the reqular sessions, a special session was set with four Invited Talks leading to the quantum manipulation technology, which is newly emerging reseach fields.

On behalf of the Organising Committee, I wish to thank the keynote speakers, invited speakers and authors of selected papers for their outstanding contributions. I would also like to thank members of the organizing committee, anonymous reviewers and volunteers for their great efforts. Without their contribution, dedication and commitment, we would not have achieved so much. Thanks also go to the Conference Secretaries (Dana Chan & Jessi Mars) for their dedicated cooperation. We sincerely hope that you will find the ICMENS 2019 beneficial and fruitful for your professional development. We also hope that you will enjoy our hospitality and will have an enjoyable and memorable time in Hiroshima.

Conference Chair, ICMENS 2019

Professor Toshikazu Ekino
Hiroshima University, Japan

IOP Conf. Series: Materials Science and Engineering **625** (2019) 011002 doi:10.1088/1757-899X/625/1/011002

Conference Committees

Advisory Committees

Prof. Richard Haverkamp, Massey University, New Zealand
Prof. C. W. Lim, City University of Hong Kong, Hong Kong

Conference Chair

Prof. Toshikazu Ekino, Hiroshima University, Japan

Local Chair

Prof. Satoshi Ishizaka, Hiroshima University, Japan

Program Chairs

Prof. Josip Brnić, University of Rijeka, Croatia
Prof. Ricardo Branco, University of Coimbra, Portugal
Prof. Syuma Yasuzuka, Hiroshima Institute of Technology, Japan

Local Committees

Prof. Noriyuki Hatakenaka, Hiroshima University, Japan
Prof. Seiji Higashitani, Hiroshima University, Japan
Prof. Masanori Inui, Hiroshima University, Japan
Prof. Norio Ogita, Hiroshima University, Japan
Prof. Akihiko Toda, Hiroshima University, Japan

Technical Committees

Assoc. Prof. Marino Brčić, University of Rijeka, Croatia
Prof. Vladimir Khovaylo, National University of Science and Technology "MISiS", Russia
Assoc. Prof. Munehiro Nishida, Hiroshima University, Japan
Asst. Prof. Masataka Iinuma, Hiroshima University, Japan
Prof. Akihiko Fujiwara, Kwansei Gakuin University, Japan
Prof. Jiin-Yuh Jang, National Cheng Kung University, Taiwan
Prof. Jui-Hsiang Liu, National Cheng Kung University, Taiwan
Prof. Ick Soo Kim, Shinshu University, Ueda, Japan
Prof. Rakesh Chand Ramola, HNB Garhwal University, India
Asst. Prof. Toshiyuki Fujii, Asahikawa Medical University, Japan
Dr. Norihito Kosugi, VINAS Co., Ltd., Japan
Dr. Abbas Mohammed Selman, University of Kufa, Iraq
Dr. Omid Nabinejad, Michigan State University, USA
Dr. Shweta Agarwala, Nanyang Technological University, Singapore
Dr. Dewi Suriyani Che Halin, Unversiti Malaysia Perlis, Malaysia
Dr. Du Ngoc Uy Lan, Unversiti Malaysia Perlis, Malaysia

IOP Conf. Series: Materials Science and Engineering **625** (2019) 011002 doi:10.1088/1757-899X/625/1/011002

IOP Conf. Series: Materials Science and Engineering **625** (2019) 011003 doi:10.1088/1757-899X/625/1/011003

Peer review statement

All papers published in this volume of *IOP Conference Series: Materials Science and Engineering* have been peer reviewed through processes administered by the proceedings Editors. Reviews were conducted by expert referees to the professional and scientific standards expected of a proceedings journal published by IOP Publishing.

IOP Conf. Series: Materials Science and Engineering **625** (2019) 012001 doi:10.1088/1757-899X/625/1/012001

Halogenated monolayer SnS: a first-principles studies

K H Yeoh[1], T L Yoon[2], Rusi[3], D S Ong[4] and T L Lim[5]

[1]Department of Electrical and Electronic Engineering, Lee Kong Chian Faculty of Engineering and Science, Universiti Tunku Abdul Rahman, 43000 Kajang, Selangor, Malaysia
[2]School of Physics, Universiti Sains Malaysia, 11800 USM, Penang, Malaysia
[3]Center for Foundation Studies, International University of Malaya-Wales, 50480 Kuala Lumpur, Malaysia.
[4]Faculty of Engineering, Multimedia University, Persiaran Multimedia, 63100 Cyberjaya, Selangor, Malaysia
[5]Faculty of Engineering and Technology, Multimedia University, Jalan Ayer Keroh Lama, 75450 Melaka, Malaysia.

Email: khyeoh@utar.edu.my, tlyoon@usm.my

Abstract. The electronic and structural properties of a halogenated monolayer SnS is investigated by using first-principles calculations. The preferred adsorption sites for F, Cl and Br are found to be on top of the Sn atom while I atom prefers to adsorb on top of the SnS ring. All the halogenated monolayers SnS exhibit metallic behavior with charge transfer from the monolayer SnS to the halogen adatom. In the relaxed system, a larger halogen atom is located further away from the monolayer SnS which results in smaller charger transfer and thus weaker binding energy. Compared to the pristine monolayer SnS, the work function of the F, Cl or Br functionalized monolayer SnS is increased by 22 %, while for the case of I adsorption, the work function is further increased by 30%.

1. Introduction

Monolayer Tin(II) Sulfide (SnS) is a phosphorene-like 2D material with puckered structure and indirect bandgap of 1.96 eV [1]. Experimentally, SnS nanosheets have been synthesized by using Liquid-Phase Exfoliation method [2]. Monolayer SnS has large adsorption coefficient, making it an excellent candidate for flexible photovoltaic (PV) applications [3]. Compared to the widely researched monolayer MoS_2, SnS possesses a higher electron mobility [4]. More importantly, the elemental composition of SnS, i.e. Sn and S are non-toxic and highly abundant. Previous theoretical studies have shown that atom adsorption can be used to functionalize 2D materials to achieve the desired physical properties [5-9]. For example, theoretical studies by Li et. al. have shown that the electronic and magnetic properties of the monolayer SnS can be controlled through the adsorption of light metals, transition metals atoms, hydrogen and oxygen [10]. Motivated by this, the theoretical studies on the interaction of a single halogen atom on a pristine monolayer SnS have been carried out in present work. The structural and electronic properties of the halogenated monolayer SnS are systematically investigated and compared against the pristine monolayer SnS.

2. Methods

Density Functional Theory (DFT) calculations are carried out using the Quantum ESPRESSO software [11,12]. The exchange correlation interactions are treated using the

Perdew–Burke–Ernzerhof (PBE) generalized gradient approximation and the Projected Augmented Wave (PAW) pseudopotential is used. The plane wave and charge density cutoff energy is set to 45 Ry and 240 Ry, respectively. For relaxation, the momentum space is sampled on a $\mathbf{8 \times 8 \times 1}$ Monkhorst-Pack grid while a denser $\mathbf{12 \times 12 \times 1}$ grid is used for the rest of the calculations. A 18 Å vacuum layer is inserted between the SnS monolayer to eliminate any possible interactions with the neighboring layers. Dipole correction and Van derWaals interactions are taken into consideration by using the DFT-D2 method [13]. All the simulated supercells are relaxed until the atomic forces are less than 0.01 eV/Å. The charge transfer is calculated based on the Löwdin charge analysis. A negative (positive) charge transfer indicate the electrons are transferred from (to) the monolayer SnS to (from) the adatom.

3. Results and discussions

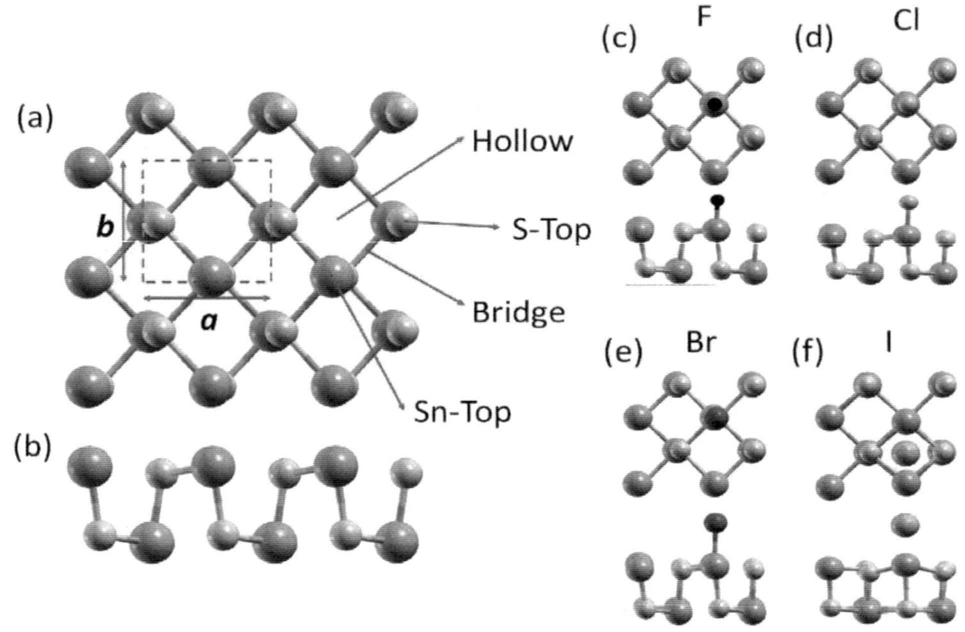

Figure 1. (a) Top view and (b) side view of a relaxed structure of the monolayer SnS. The red-dotted box indicates the unit cell with lattice constant $\boldsymbol{a}$ and $\boldsymbol{b}$. The preferred adsorption site for F, Cl, Br and I is shown in (c), (d), (e) and (f), respectively.

In present work the adsorption of single halogen atom on a $\mathbf{2 \times 2}$ supercell which corresponds to the adatom coverage of $\Theta = 1/16$ are being considered. The calculated value of lattice constant $\boldsymbol{a}$ and $\boldsymbol{b}$ of the pristine monolayer SnS are 4.25 Å and 4.08 Å, respectively, which is in close agreement with those reported in [9]. There are four possible adsorption sites for the halogen atom as shown in figure 1 (a) i.e. on top of the SnS ring (Hollow), above the S atom (S-Top), above the Sn atom (Sn-Top) and on top of the Sn-S bridge (Bridge) [9]. A single halogen atom is placed on these adsorption sites and variable cell relaxation is carried out. The most preferred adsorption sites are those with the lowest binding energy. As shown in figure 1 (c), (d) and (e), F, Cl and Br prefer to adsorb on the Sn-Top whereas I preferred to adsorb on the Hollow site. The binding energy of the halogen functionalized monolayer SnS is calculated as:

$$E_B = E_{SnS+adatom} - E_{SnS} - E_{adatom} \quad (1)$$

where E_{adatom} is the energy of the isolated halogen adatom, E_{SnS} is the total energy of the pristine monolayer SnS supercell and $E_{SnS+adatom}$ is the total energy of the monolayer SnS supercell adsorbed with halogen adatom.

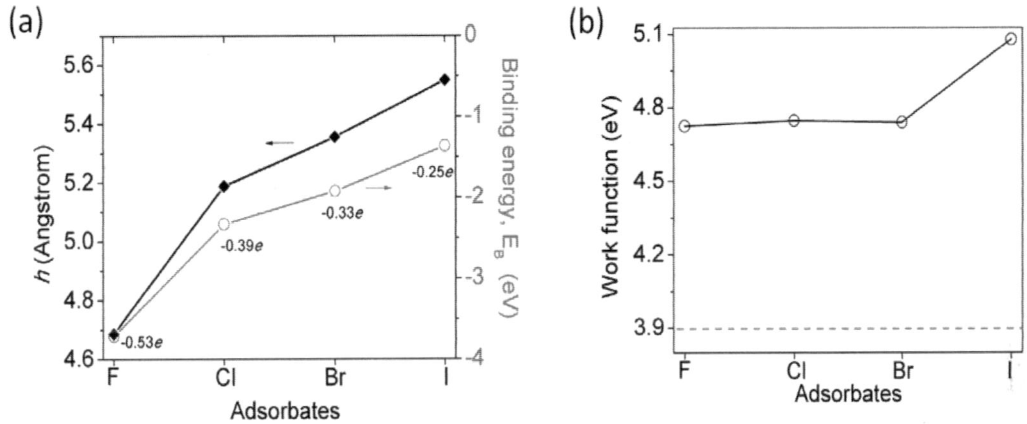

Figure 2. (a) The average distance of the halogen adatom from the bottom lying monolayer SnS layer, $\boldsymbol{h}$ and binding energy, $E_B$of the functionalized monolayer SnS for different types of halogen adatoms. The numerical values near the red hollow symbols denote the value of the charge transfer. (b) The calculated work function of the halogenated monolayer SnS. The red-dotted line indicates the work function of a pristine monolayer SnS.

As shown in figure 2 (a), all the E_B has a negative value, indicating the halogenated monolayer SnS is energetically stable. As the size of the halogen atom increases from F to I, the adatom is located further away from the monolayer SnS. This is evident by the larger $\boldsymbol{h}$ as illustrated in figure 2 (a). Following the same trend, $|E_B|$ reduces when the atomic mass of the halogen atom increases, indicating a weaker bond between the halogen atom and monolayer SnS. For example, the $|E_B|$ for the case of F adsorption is 3.72 eV and reduces gradually to 1.36 eV when the halogen atom change to I. It is worth mentioning that similar trend is observed for halogenated borophene which was reported by Khanifaev et. al. recently [14]. To gain further insight into the interaction of the halogen adatoms on the monolayer SnS, the charge transfer of the adatoms are calculated. With the adsorption of F, 0.53e is transferred from the monolayer SnS into the F adatom. However, the charge transfer is reduced by more than **2** × for the case of I adsorbed monolayer SnS. In fact, the charge transfer reduces when the atomic mass of the halogen atom increases. A higher charge transfer could provide an explanation for stronger $|E_B|$ as can be observed in figure 2 (a). The direction of charge transfer for Cl and Br is similar to F and I. This implies that the halogenated monolayer SnS is hole rich. As shown in figure 2 (b), the work function of the pristine monolayer SnS is predicted to be 3.9 eV. The work function is increased for the halogenated monolayer SnS. This is expected because halogen atoms are known to be strong electron acceptors. With the adsorption of F. Cl or Br, the work function of the monolayer SnS is increased by about 22 %. On the other hand, when the monolayer SnS adsorbed with I, its work function is further increased to 5.08 eV, which corresponds to 30% increment compared to the pristine monolayer SnS. Figure 3 (a), (b), (c) and (d) show the band structures, density of states (DOS) and projected density of states (PDOS) of the halogenated monolayer SnS. F, Cl or Br adsorption results in the formation of the gap states with the Fermi level located in those localized energy states. These gap states are contributed by the p- states of the adsorbates. In contrast, for the case of I, localized energy states are not observed. Unlike pristine monolayer SnS which is a semiconductor, all halogenated monolayer SnS exhibits metallic characteristics as evident by the presence of Fermi level in the energy states.

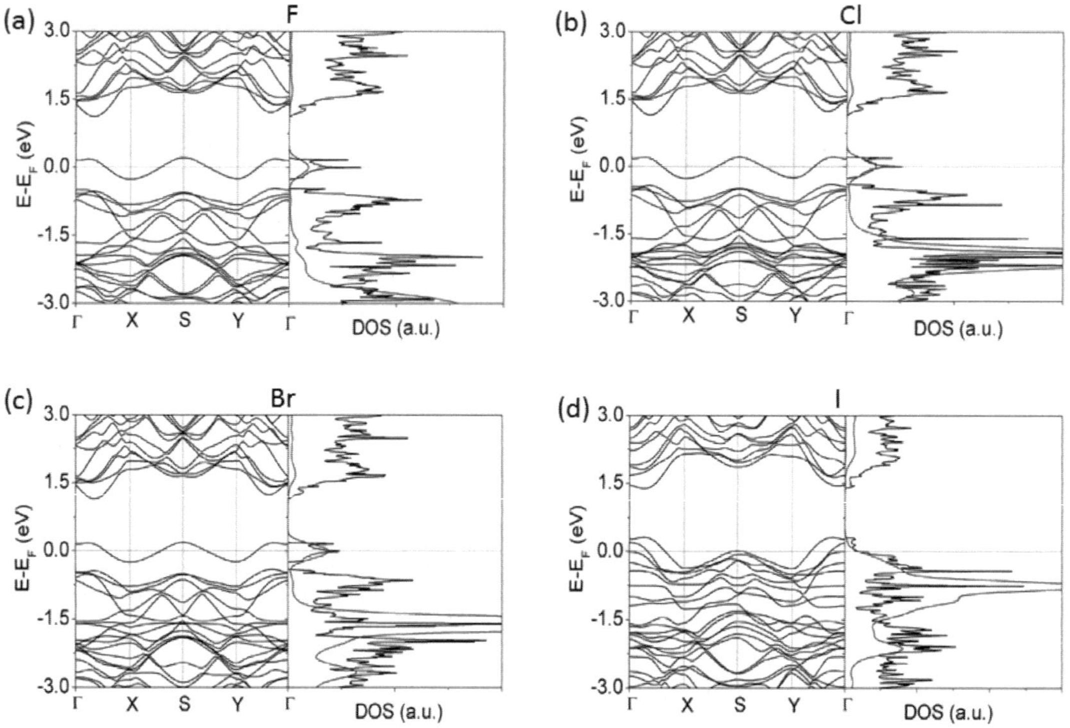

Figure 3. Electronic band structure, density of states (DOS) and partial density of states (PDOS) of monolayer SnS functionalized with (a) F, (b) Cl, (c) Br and (d) I. The red and blue colour line indicates the projected states of the halogen adatom for *s* and *p* orbital respectively. The PDOS is magnified by $\mathbf{10}\times$ for better viewing convenience.

4. Conclusions

By using first-principles calculations, a systematic investigation on the structural and electronics properties of a monolayer SnS functionalized with halogen atoms have been carried out. No significant structural deformation is observed with the adsorption of halogen atom on the monolayer SnS. The $\boldsymbol{E_B}$ of the halogen atom is strongly correlated with the charge transfer, the distance of the halogen atom from the monolayer SnS and the size of the halogen atom. Upon the adsorption of halogen atom, larger halogen atom is located further away from the monolayer SnS and subsequently has lower charge transfer. This results in a weaker binding energy. All halogenated monolayer SnS exhibit metallic characteristic and it is shown the work function can be increased up to 30% compared to the pristine monolayer SnS. Present work can be extended by simulating larger supercell thus enable us to investigate the effects of the halogen concentration adsorbed on the monolayer SnS. Present results could provide further fundamental understanding on the physical properties of halogenated monolayer SnS.

Acknowledgement
This work is partially supported by MOHE under grant number FRGS/1/2017/STG07/UTAR/02/2. YTL acknowledges the financial support of USM Bridging (2018).

IOP Conf. Series: Materials Science and Engineering **625** (2019) 012001 doi:10.1088/1757-899X/625/1/012001

References

[1] Xin C, Zheng J, Su Y, Li S, Zhang B, Feng Y and Pan F. 2016 *J. Phys. Chem.* C **120** 22663
[2] Brent J R, Lewis D J, Lorenz T, Lewis E A, Savjani N, Haigh S J, Seifert G, Derby B and O'Brien P 2015 *J. Am. Chem. Soc.* **137** 12689
[3] Zhang Y, Shang B, Li L and Lei J 2017 *RSC Adv.* **7** 30327
[4] Wu M and Zeng X C 2016 Nano Lett. 16 3236
[5] Yeoh K H, Yoon T L, Ong D S, Lim T L and Abdullahi Z 2017 *Phys Chem Chem Phys.* **19** 25786.
[6] Yeoh K H, Yoon T L, Kumaresan H A, Chew K –H, Ong D S, Lim T L and Rusi 2018 *J. Phys. Chem. Solids.* **123** 294
[7] Gao N, Zheng W T and Jiang Q 2012 *Phys. Chem. Chem. Phys.***14** 257.
[8] Ma Y, Dai Y, Niu C and Huang B 2012 *J. Mater. Chem.* **22** 12587.
[9] Li Y, Xia C, Du J, Xiong W, Li X and Wei S 2017 *Phys.Chem.Chem.Phys.* **19** 5423
[10] Sahin H and Peeters F M 2013 *Phys. Rev. B.* **87** 085423.
[11] Giannozzi P, et al. *2009 J. Phys.: Condens. Matter.* **21** 395502
[12] Giannozzi P, et al. *2017 J. Phys.: Condens. Matter.* **29** 465901
[13] Grimme S. 2006 *J. Comput. Chem.* **27** 1787.
[14] Khanifaev J, Peköz R, Konuk M and Durgun E. 2017 *Phys.Chem.Chem.Phys.* **19** 28963

IOP Conf. Series: Materials Science and Engineering **625** (2019) 012002 doi:10.1088/1757-899X/625/1/012002

Silicon-doped indium oxide – a promising amorphous oxide semiconductor material for thin-film transistor fabricated by spin coating method

Ha Hoang[1], Kazutaka Sasaki[1], Tatsuki Hori[1], Kazuhito Tsukagoshi[2], Toshihide Nabatame[3], Bui Nguyen Quoc Trinh[4,5] and Akihiko Fujiwara[1,*]

[1]School of Science and Technology, Kwansei Gakuin University, 2-1 Gakuen, Sanda, Hyogo 669-1337, Japan
[2]National Institute for Materials Science (NIMS), International Center for Materials Nanoarchitectonics (WPI-MANA), 1-1 Namiki, Tsukuba, Ibaraki 305-0044, Japan
[3]National Institute for Materials Science (NIMS), MANA Foundry and MANA Advanced Device Materials Group, 1-1 Namiki, Tsukuba, Ibaraki 305-0044, Japan
[4]Vietnam National University – Hanoi, University of Engineering and Technology, 144 Xuan Thuy, Cau Giay, Hanoi, Vietnam
[5]Vietnam National University – Hanoi, Vietnam Japan University, Luu Huu Phuoc, Nam Tu Liem, Hanoi, Vietnam

*Email: akihiko.fujiwara@kwansei.ac.jp

Abstract. Silicon-doped indium oxide (In-Si-O or ISO) has been investigated as the channel material of thin-film transistor (TFT) for the application of next generation flat panel displays. Because solution processing is simple, low cost and low power consumption in comparison with physical vapour deposition, it is a potential candidate for TFTs fabrication. We have been exploring research on TFT using ISO system via spin coating method. In this work, the performance of 3 at.% Si-doped indium oxide TFT has been improved with the highest mobility of 3.8 cm^2/Vs without the passivation layer.

1. Introduction

For the application of flat panel displays, single crystal silicon has been often used as the channel material of thin-film transistor (TFT) because of high performance although it is expensive and lack of flexibility. Poly crystal silicon is less expensive; however, the existence of grain boundaries reduce the carrier mobilities. Amorphous silicon is the low cost and flexible, but the performance was the worst because the covalent bond sp^3 σ states dramatically reduced carrier mobilities in amorphous phase [1]. Therefore, conventional amorphous silicon was unable to apply for thin-film transistor in application of high precision flat panel display.

Recently, amorphous In-Ga-Zn-O system was reported by Prof. Hosono group with great mobility and high transparency [2-5]. Thanks to that, amorphous doped indium oxide materials have been investigated as the channel material of thin-film transistor in application of flat panel display. Among many doped elements for indium oxide system, Si is one of the high rated element because the energy bonding of Si-O is stronger than that of In-O: the Si-O bond dissociation energy is 799 kJ mol^{-1} while that of In-O is 346 kJ mol^{-1} [6]. Hence, in silicon-doped indium oxide (In-Si-O or ISO) system, a small amount of Si can significantly contribute to decrease oxygen deficiency phenomenon, stabilizing the structure of indium oxide [7-12].

ISO TFTs have been fabricated by physical vapour deposition and exhibited high mobility around 10 to 20 cm^2/Vs [9-11]. Compared to physical vapour deposition, spin coating process is a promising method for fabrication of TFT because of its simplicity, low cost and low power consumption [13-19]. Especially, it is easy to change the concentration of Si doping to characterize the Si doping effect and can be applied in large area. For those above reasons, recently we have first investigated ISO TFT via spin coating method [20-21]. In this work, the performance of ISO TFT has been improved to satisfy the operation.

2. Experimental method

The ISO solution was prepared by dissolving indium chloride powder into acetonitrile solvent with molarity of 0.05 M. A small amount of ethylene glycol was added as a minor solvent in order to achieve the uniform film with the volume ratio of ethylene glycol per acetonitrile was 1/50. It is because acetonitrile has a low boiling temperature, evaporating quickly during spin coating process; as a result, the film could not be deposited well. Hence, the existence of high boiling temperature solvent (ethylene glycol in this case) plays an important role in the formation of film [22]. The solution made by the mixture of low boiling temperature solvent (main solvent) and high boiling temperature (minor solvent) is popular in spin coating method. Tetraethyl orthosilicate was added into prepared solution with the molar ratio of 3% as for Si-doping purpose. After that, the solution was stirred by a magnetic stirrer to dissolve indium chloride powder well in the solvent of acetonitrile, ethylene glycol and tetraethyl orthosilicate.

A heavily doped p-type single crystal (100) wafer with 250 nm thickness of thermally oxidized SiO_2 layer on the surface was purchased from Electronics and Materials Corporation Limited, and used in this work as a substrate. The substrate was sonicated in acetone of 10 minutes, and dipped in sodium hydroxide 1 M of 5 minutes to modify the surface property of substrate from hydrophobic to hydrophilic via the attachment of the functional group -OH. Because the prepared ISO solution was a polar solution, so to obtain the wetting contact between substrate and solution, hydrophilic surface of substrate is essential. Then the solution was dropped on the prepared substrate and spin coated with the speed of 3000 rpm in 30 seconds. After that, the sample was dried at 100 °C by a hot plate in 5 minutes. The spin coating and drying process were repeated 4 times to achieve the desired thickness around 10 nm (according to x-ray reflectivity measurement). Finally, the samples were annealed at different temperature from 300 to 600 °C (step of 100 °C) in atmosphere in 1 hour.

To fabricate the thin-film transistor, 200-nm-Al-source electrode and drain electrode were deposited by thermal evaporation through a stencil mask. The mask was designed with various channel length from 50 to 350 μm (with a 50-μm step). The structure of ISO TFT and the fabricated ISO TFT are illustrated in Figure 1(a) and (b). The transfer characteristic (the drain current I_D versus the gate-source voltage V_{GS}) and output characteristic (the drain current I_D versus the drain-source voltage V_{DS}) of ISO TFT were indicated by a Keysight B2912A system. Because the applied V_{GS} was up to 80 V, so the mobility was estimated in the saturation regime as following equation:

$$\mu = \frac{2L}{WC_i}\left(\frac{\partial\sqrt{I_D}}{\partial V_{GS}}\right)^2 \tag{1}$$

where L and W are the channel length and width, C_i is the capacitance per unit area of the gate insulator, and V_{GS} is the gate-source voltage [23-24].

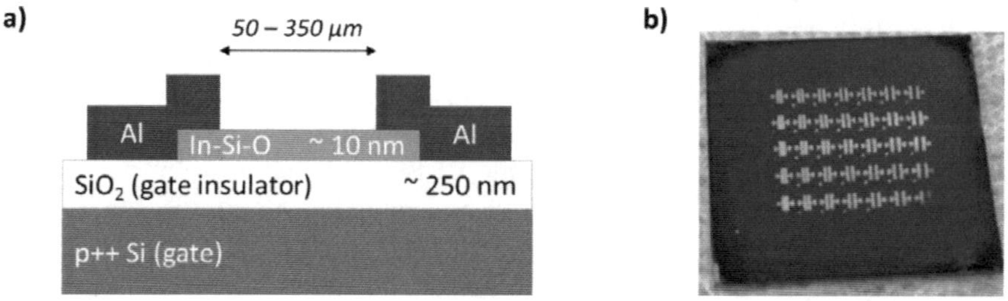

Figure 1. (a) The cross-section structure of the ISO TFT. (b) The fabricated ISO TFT with 7 different channel lengths between 50 and 350 μm (a 50-μm step) based on the view from left to right side.

3. Results and discussion

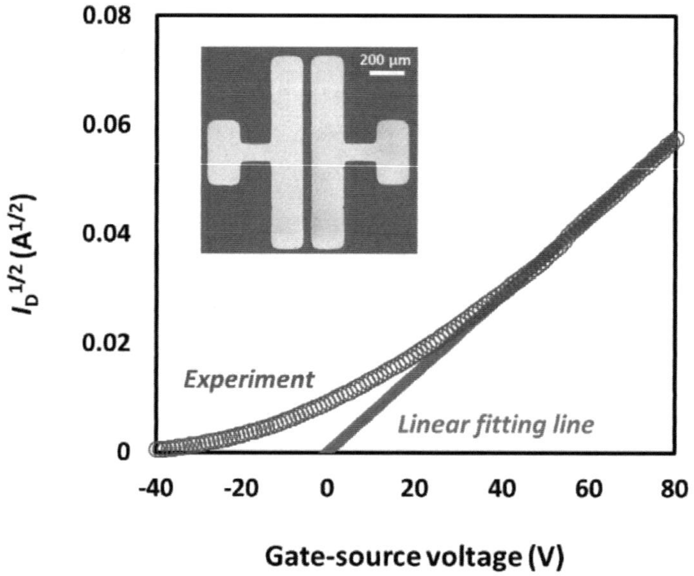

Figure 2. The plot of $I_D^{1/2}$ versus the gate-source voltage V_G of the best performance ISO (3 at.% Si) TFT annealed at 400 °C in air with $L = 50$ μm and $W = 1000$ μm. The circle mark is the raw data from experiment while the solid line is the linear fitting line. The figure includes the microscope image of the ISO TFT.

In our recent studies, x-ray diffraction, x-ray reflectivity and thermal desorption spectroscopy measurements revealed that ISO thin films fabricated by this procedure were amorphous form without organic residues, and their thickness, density, RMS roughness were about 10 nm, about 6 g/cm^3 and less than 1 nm, respectively [12,20,21]. Besides, the ISO TFTs were examined under numerous conditions such as ageing effect, Si-doping effect, the dependent on annealing temperature and the dependent on the film thickness for optimization [21]. It is suggested that the optimal fabrication involved the annealing temperature of 400 °C in atmosphere in 1 hour. In this work, in order to eliminate extrinsic effect reducing the device performance, by repeating the fabrication process many times in clean room and shorten the gap time between NaOH treatment and spin coating step, at the moment, the highest mobility we have achieved is approximately 3.8 cm^2/Vs without the passivation layer as shown in Figure 2. Here, the applied gate-source voltage increased from -40 to 80 V while the drain-source voltage was 40 V: the measurements were conducted at the saturation region. From the square root of I_D, the slope of the fitting line was used to estimate the saturated mobility based on equation (1) and the intercept of the fitting line with x-axis was defined as the value of the threshold voltage. The result showed that the threshold voltage was around 0 V although there was an enhancement of I_D in the experimental data around $V_{GS} = 0$ V. It is due to the traps at the interface of substrate and ISO films. To solve this problem, the substrate should be cleaned well by additional processes to remove contaminants completely.

As the report by Han *et al*, the presence of ozone in annealing process not only contributed to high performance of TFT but also reduced the annealing temperature [22]. As the next work for further reduction of oxygen vacancy, we also consider to anneal the ISO thin films in ozone or mixture of ozone/oxygen environment to increase the performance of solution-processed ISO TFT.

4. Conclusions

By improving fabrication process of our recent work in which the 3 at.% Si-doped indium oxide thin films were manufactured by the spin coating technique, investigated on various fabrication condition, and the best mobility was exceed 1 cm^2/Vs, we achieved the highest mobility up to 3.8 cm^2/Vs. Further work will be continued to increase the performance of solution-processed ISO TFTs for high precision displays.

Acknowledgements

In this work, we would like to thank to the Hyogo Overseas Research Network (HORN) Program, Hyogo Earthquake Memorial 21st Century Research Institute, and Grants-in-Aid for Scientific Research (grant No. JP15H03568). Also, we would like to acknowledge the research project supported in 2019, by Vietnam Japan University (VJU) Research Grant Program of Japan International Cooperation Agency (JICA).

References

[1] Yu X, Marks T J and Facchetti A 2016 *Nature* Materials **15** 383–396
[2] Nomura K, Ohta H, Takagi A, Kamiya T, Hirano M and Hosono H 2004 *Nature* **432** 488–492
[3] Kamiya T, Nomura K and Hosono H 2010 *Sci. Technol. Adv. Mater.* **11** 044305
[4] Kamiya T and Hosono H 2010 *NPG Asia Mater.* **2** 15–22
[5] Nomura K, Kamiya T, Ikenaga E, Yanagi H, Kobayashi K and Hosono H 2013 *J. Appl. Phys.* **114** 163713
[6] Luo Y R 2009 *CRC Handbook of Chemistry and Physics* (CRC Press/Taylor and Francis)
[7] Mitoma N, Aikawa S, Gao X, Kizu T, Shimizu M, Lin M F, Nabatame T and Tsukagoshi K 2014 *Appl. Phys. Lett.* **104** 102103
[8] Mitoma N, Aikawa S, Yang W O, Gao X, Kizu T, Lin M F, Fujiwara A, Nabatame T and Tsukagoshi K 2015 *Appl. Phys. Lett.* **106** 042106
[9] Aikawa S, Mitoma N, Kizu T, Nabatame T and Tsukagoshi K 2015 *Appl. Phys. Lett.* **106** 192103
[10] Kizu T, Aikawa S, Nabatame T, Fujiwara A, Ito K, Takahashi M and Tsukagoshi K 2016 *J. Appl. Phys.* **120** 045702
[11] Aikawa S, Nabatame T and Tsukagoshi K 2013 *Appl. Phys. Lett.* **103** 172105
[12] Jan H E, Hoang H, Nakamura T, Koga T, Ina T, Uruga T, Kizu T, Tsukagoshi K, Nabatame T and Fujiwara A 2017 *J. Electron. Mater.* **46** 3610–3614
[13] Yu X et al 2015 *Proc. Natl. Acad. Sci. USA* **112** 3217–3222
[14] Smith J, Zeng L, Khanal R, Stallings K, Facchetti A, Medvedeva J E, Bedzyk M J and Marks T J 2015 *Adv. Electron. Mater.* **1** 1500146
[15] Park J H, Yoo Y B, Lee K H, Jang W S, Oh J Y, Chae S S, Lee H W, Han S W and Baik H K 2013 *ACS Appl. Mater. Interfaces* **5** 8067–8075

IOP Conf. Series: Materials Science and Engineering **625** (2019) 012002 doi:10.1088/1757-899X/625/1/012002

[16] Jeong S, Jeong Y and Moon J 2008 *J. Phys. Chem. C* **112** 11082–11085
[17] Norris B J, Anderson J, Wager J F and Keszler D A 2003 *J. Phys. D: Appl. Phys.* **36** L105
[18] Niederberger M, Gamweitner G, Buha J, Polleux J, Ba J and Pinna N 2006 *J. Sol-Gel Sci. Technol.* **40** 259–266
[19] Lee D H, Chang Y J, Herman G S and Chang C H 2007 *Adv. Mater.* **19** 843–847
[20] Hoang H, Hori T, Yasuda T, Kizu T, Tsukagoshi K, Nabatame T, Bui T and Fujiwara A 2018 *25th International Workshop on Active-Matrix Flatpanel Displays and Devices* (IEEE) (DOI: 10.23919/AM-FPD.2018.8437420)
[21] Hoang H, Hori T, Yasuda T, Kizu T, Tsukagoshi K, Nabatame T, Bui T and Fujiwara A 2019 *Mater. Res. Express* **6** 026410
[22] Han S Y, Herman G S and Chang C H 2011 *J. Am. Chem. Soc.* **133** 5166–5169
[23] Sze S M and Lee M K 2012 *Semiconductor Device: Physics and Technology* (Wiley)
[24] Matsuoka Y, Uno K, Takahashi N, Maeda A, Inami N, Shikoh E, Yamamoto Y, Hori H and Fujiwara A 2006 *Appl. Phys. Lett.* **89** 173510

IOP Conf. Series: Materials Science and Engineering **625** (2019) 012003 doi:10.1088/1757-899X/625/1/012003

Analysis of the mechanical response of materials used in design for highly stressed components

Josip Brnic[1,*], **Sanjin Krscanski**[1] **and Marino Brcic**[1]

[1] University of Rijeka, Faculty of Engineering, Vukovarska 58, 51000 Rijeka, Croatia

[*]Email: brnic@riteh.hr

Abstract. In this paper both the mechanical properties and responses of materials usually used in engineering for design and manufacture of highly stressed components were considered. Between the materials used for the aforementioned purposes are considered the following steel alloys: 30CrNiMo8, 20MnCr5, 51CrV4. Investigated mechanical properties of the materials were ultimate tensile strength, yield strength and modulus of elasticity, while in the area of mechanical responses were explored / tested creep behaviour and fatigue of the considered materials. In general, all of the responses are presented in the form of stress-strain curves, creep curves and S-N curves (maximum applied stress versus the number of the cycles to failure). Mechanical properties were tested at different temperatures, creep behaviour was investigated at different temperatures when material was subjected to stresses at prescribed percentage of yield strength while fatigue of the considered materials was investigated on smooth unnoched specimen for prescribed maximum uniaxial stress and for prescribed stress ratio. In addition, all of obtained data were analysed and compared regarding the type of used materials.

1. Introduction

Different types of materials are used in different environmental conditions. This means that material must be chosen in accordance with the purpose of the structure, the environmental operating conditions and with the choice of the process by which it is to be shaped or joined [1]. Under structure, it is understood any type of construction or machine. In the considered case, used material must be able to take high stresses, dynamic loads, and must be creep resistant. On the basis of this, follows, for example, that material can be exposed to high stresses, high temperatures, creep and fatigue. Regarding the quality and reliability of the structure, taking into consideration the entire structure life from design through manufacture and exploitation, several key problems must be analysed. These problems are: purpose for which structure is intended to be used, environmental operating conditions, choice of material, lifetime prediction, safety, common causes of failures, possible (common) failure modes, analyses of how and why structural component has failed, etc. Causes of the failures and appropriate failure modes are interconnected. In engineering practice, structure is usually designed and manufactured in order to guarantee that it does not contain any failure. However, this does not have to be the correct approach. Common causes of failures are [2- 3]: inadequate choose of material, design errors (used material, the size and shape of an engineering component, properties), manufacturing defects, structural loading, temperature effects, misuse (the structure is subjected to conditions for which it was not designed), assembly errors, improper maintenance, unforeseen operating conditions, inadequate control, etc. These causes may be divided into pre-existing causes and that appearing in service life. The same applies to failure modes [4]. Commonly observed failures mode in engineering practice are [5]: yielding, creep, fatigue, wear, corrosion, impact, buckling, etc. Any of particular mechanical failure has its cause of origin and the form of expression (mode of representation), i.e., failure mode. In this sense the cause can provide the answer why an engineering component has failed while the failure mode enables the answer how an engineering component has failed. In addition, it is useful to mention that science categories are involved in this problematic. In this sense, materials science primarily concerned with the basic knowledge of materials while materials engineering concerned with the use of the knowledge of materials. In accordance with the considered problems within this paper, i.e., design of highly stressed engineering components, the following subjects related to the applied materials were explored and analysed: mechanical properties, creep and fatigue of material. Designer who deals with material selection must be familiar with all of mentioned subjects. Some details of investigated responses of considered materials can be found in the literature. In this sense, in Ref. [6], effects of welding parameters related to 30CrNiMo8 steel were investigated, while, related to the same material, an analysis the relationship between cutting forces and tool wear during turning is presented in Ref. [7]. As far as 20MnCr5 steel is concerned, it is suggested to review papers published in Refs. [8-9]. In Ref. [8] material structure changes developing in the bore of gears made of a material of 20MnCr5 is considered, while in Ref. [9] dependence of material properties versus temperature are investigated. In addition, in Ref. [10] the microstructural features of 51CrV4 alloy, used as spring steel component, and its fatigue behavior was investigated. In Ref. [11] some of mechanical properties related to 51CrV4 steel and its responses are investigated. Three considered materials are selected for testing with respect to their use in high-stressed engineering elements, as above mentioned.

2. Data related to the tests

Chemical compositions of investigated materials are shown in table 1. All of tested materials can be used in applications for highly stressed engineering components. The equipment used in testing procedures were testing machine Zwick/Roell (400 kN), macro-extensometer, high temperature extensometer (900°C), dynamic testing machine (± 50 kN). Testing procedures were performed in accordance with the standards contained in [12]: ASTM: E8M-15a (tensile testing at room temperature), ASTM: E21-09 (tensile testing at high temperatures), ASTM: E 139-11 (creep testing) as well as ISO standard 12017:2012 (tensile fatigue testing), [13].

Table 1. Tested materials: Chemical compositions (some elements only), mass %

Material / round bar (mm)	Chemical composition (some elements only), mass %					
	C	Cr	Si	Ni	Mn	Mo
30CrNiMo8 (1.6580/AISI 4340) /20	0.29	2.07	0.31	1.89	0.41	0.24
20MnCr5 (1.7147/AISI 5120) /18	0.22	1.11	0.29	0.08	1.23	0.01
51CrV4 (1.8159/AISI 6150) /20	0.45	1.09	0.32	0.22	0.92	0.005

Delivered state of the material and its description:

30CrNiMo8 (annealed, low alloy); 20MnCr5 (annealed, low-carbon, low chromium); 51CrV4 (annealed, chromium-vanadium). All of tested specimens were manufactured from round bars according to the appropriate standards. In this sense, specimens used in the procedure of mechanical properties determination were manufactured in accordance with ASTM standards [12], while those used in fatigue testing (smooth unnotched specimen) in accordance with ISO standard [13]. In each type of test, 3-5 specimens were tested. The results presented mostly refer to the first test. All tests were performed as uniaxial tests, and at fatigue testing procedure was conducted as stress-controlled test. With respect to the geometry of the specimen no stress concentration factor was considered, and also, no multiaxial stress state can be addressed since no multiaxial loading was considered.

3. Results of research and analysis

3.1 Determination of mechanical properties based on engineering stress-strain diagrams

Engineering stress-strain diagrams at different temperatures for tested materials are shown in figure 1.

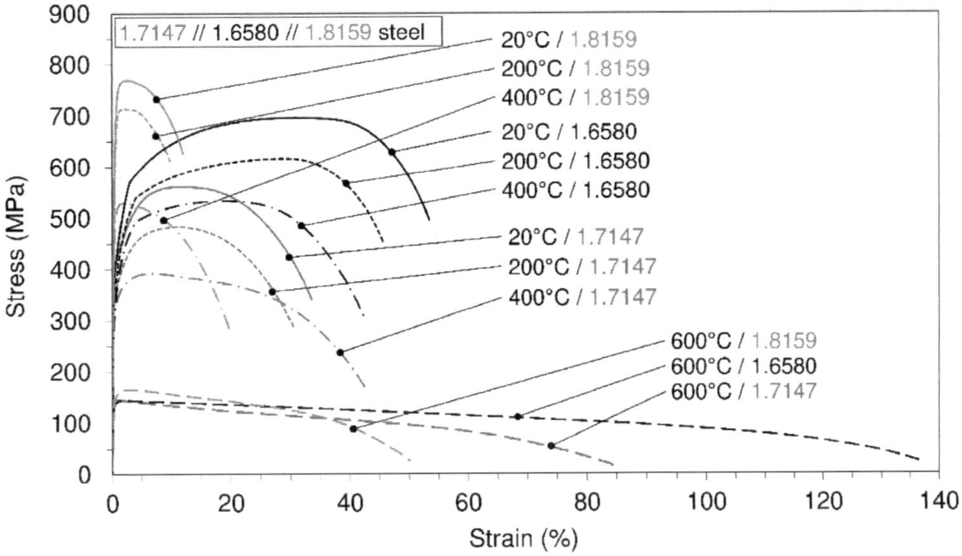

Figure 1. Engineering stress-strain diagrams at different temperatures: 1.8159 (51CrV4), 1.6580 (30CrNiMo8), 1.7147 (20MnCr5).

Mechanical properties as well as elongations and contractions can be derived on the basis of above presented diagrams. As it is visible 1.8159 steel has the highest values of the mechanical properties and the lowest elongation at the same time.

3.2. Creep test results

Short-time creep tests performed at different temperatures and different stress levels were considered. These tests show the mechanical behavior of the material during considered time, i.e., resistance of material to creep. Creep is usually defined as time dependent process showing continuously material strain increase while the stress (load) is kept constant [14]. It is appreciable at temperature above 0.4 of melting temperature [15] and in engineering practice it is usually allowed within the limits (1-2) %. For each of the tested materials creep behavior is presented in figure 2. All of materials were tested at the same temperatures and very similar stress levels regarding the percentage of their yield strength. As it is visible, at temperatures of 400°C steel 1.8159 has the parabolic shape for the longest period. In accordance with this, at higher temperatures and at new selected stress levels, this type of steel changes its shape of creep curve the slowest. From this point of view, it is apparent which material can be treated as the most resistant to creep.

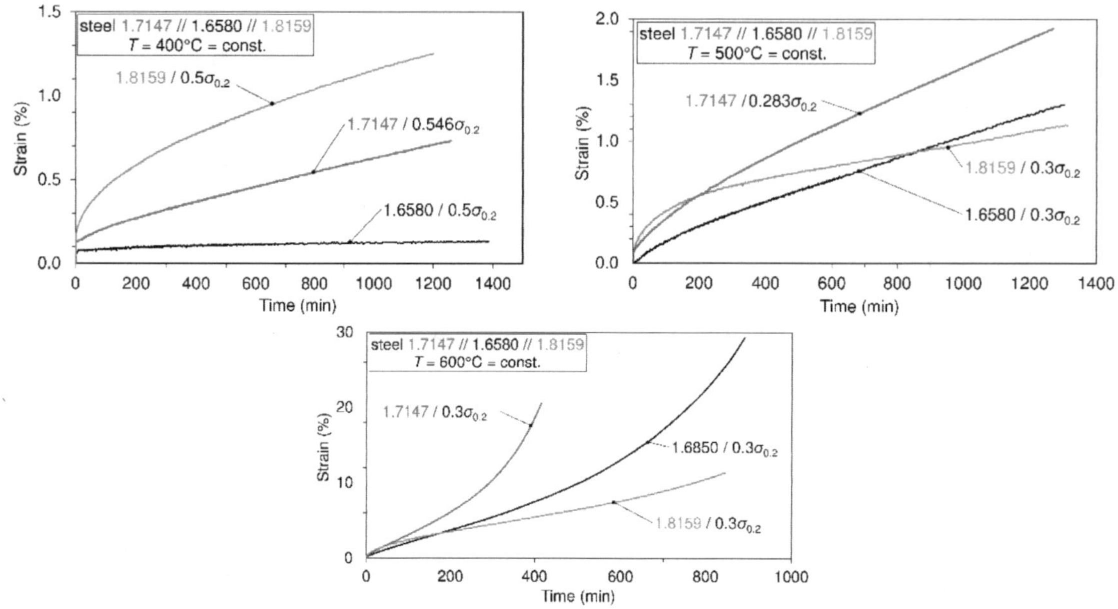

Figure 2. Creep behavior of the tested materials at different temperatures and selected stress levels: 30CrNiMo8 (1.6580), 20MnCr5 (1.7147), 51CrV4 (1.8159)

3.3 Fatigue limit determination based on fatigue testing

Fatigue is one of very often encountered failure modes in dynamically stressed machine parts. Experience testifies that fracture of machine parts are often due to fatigue during regular operating conditions [16]. Based on fatigue tests performed on the specimens subjected to uniaxial cyclic loading at stress ratio R = −1, for some of tested materials, using ISO 12107:2012 (2012) standard, fatigue limit (endurance limit) was determined. Results of the fatigue tests are presented in figure 3 where on the ordinate (σ_{max}/MPa) data related to the maximum stresses is plotted, while on the abscissa (N) data related to the number of the cycles to failure is plotted. This is so called S – N diagram (Wohler curve or fatigue life diagram). It is visible that steel 1.8159 (51CrV4) has higher fatigue limit than steel 1.6580 (30CrNiMo8). There was not enough test specimens to test the fatigue of the steel 1.7147 (20MnCr5).

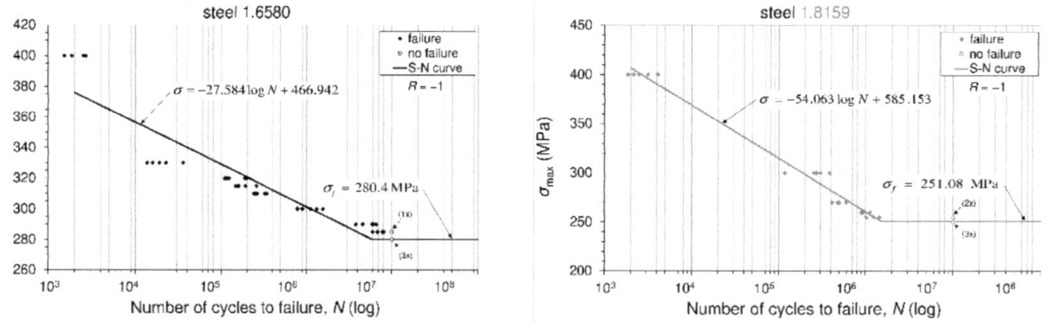

Figure 3. Stress versus number of the cycles to failure: S – N curves.

4. Conclusion

The most important properties for all of investigated materials were determined. In this sense, at room temperature, steel 51CrV4 has the highest both ultimate tensile strength and yield strength (770 MPa, 642 MPa), while 30CrNiMo8 has (696 MPa, 583 MPa) and 20MnCr5 has (561 MPa, 397 MPa). Steel 51CrV4 can also be considered as the most effective regarding the creep resistance. In accordance with experimental results steel 30CrNiMo8 (1.6580) exhibits the best fatigue limit. However, all of considered steels are suitable for use in design for highly stressed engineering components.

Acknowledgments
This work has been supported by the University of Rijeka within the projects: uniri-technic-18-42, “Investigation, analysis and modeling the behavior of structural elements stressed at room and high temperatures” and uniri-technic-18-200, “Failure analysis of materials in marine environment”

References

[1] Ashby M F 2011 *Materials Selection in Mechanical Design* (Burlington: Butterworth-Heinemann / Elsevier Ltd)
[2] Brnic J, Turkalj G, Niu J, Canadija M and Lanc D 2013 Analysis of experimental data on the behavior of steel S275JR – reliability of modern design Materials & Design **47** 497-504.
[3] Brooks C R and Choundury A 2002 *Failure Analysis of Engineering Materials* (USA: McGraw-Hill)
[4] Brnic J 2018 *Analysis of Engineering Structures and Material Behavior* (Chichester: John Wiley & Sons)
[5] Collins J A 1993 *Failure of Materials in Mechanical Design* 2nd edition (New York: John Wiley & Sons)
[6] Effertz P D S, Fuchs F and Enzinger N 2018 The influence of process parameters in linear friction welded 30CrNiMo8 small cross-section: a modelling approach *Science and Technology of Welding and Joining* 1-9 Article in press.
[7] Anicic O, Jovic S, Stanojevic N, Marsenic M, Pejovic B and Nedic M 2018 Estimation of tool wear according to cutting forces during machining procedure *Sensor Review* **38** 2 176-180.
[8] Kundrak J, Gacsi Z, Gvani K, Bana V and Tomolva G 2012 X-ray diffraction investigation of white layer development in hard-turned surface *International Journal of Advanced Manufacturing Technology* **62** 5-8 457-69.
[9] Brnic J, Turkalj G, Lanc D, Canadija M, Brcic M and Vukelic G 2014 Comparison of material properties: steel 20MnCr5 and similar steels *Journal of Constructional Steel Research* **95** 81-9.
[10] Kömeç A, Dikçi K, Atapek S H, Polat Ş and Çelik G A 2017 Microstructural and mechanical characterization of the parabolic spring steel 51CrV4 *Materialpruefung/Materials Testing* **59** 11-12 997-1002.
[11] Brnic J, Brcic M, Krscanski S, Lanc D Niu J and Wang P 2018 Steel 51CrV4 under high temperatures, short- time creep and high cycle fatigue, *Journal of Constructional Steel Research*, **147** 468-476.
[12] *Metallography: Metals–Mechanical Testing: Elevated and Low–Temperature Tests (Annual Book of ASTM Standards vol. 03.01)* 2015 (ASTM International: Baltimore)
[13] *12107:2012-Metallic Materials-Fatigue Testing-Statistical Planning and Analysis of Data 2012 (Geneva: International Organization for Standardization)*
[14] Solecki R, Conant P R. 2003 *Advanced Mechanics of Materials*, Oxford University Press (New York).
[15] Raghavan V 2004 *Materials Science and Engineering* (New Delhi: Prentice- Hall of India).
[16] Bathias C and Pineau A 2010 *Fatigue f Materials and Structures* (UK and USA: ISTE Ltd and John Wiley & Sons)

IOP Conf. Series: Materials Science and Engineering **625** (2019) 012004 doi:10.1088/1757-899X/625/1/012004

Equivalent beam model of single walled carbon nanotube with imperfections

Marino Brčić[1], Marko Čanađija[1] and Josip Brnić[1]

[1]University of Rijeka, Faculty of Engineering, Vukovarska 58, HR-51000 Rijeka, Croatia

Email: mbrcic@riteh.hr

Abstract. The presented paper is a continuation to authors' previous work on numerical modelling of carbon nanotubes (CNTs) and CNT reinforced nanocomposite materials. One of the numerical method approach for CNT modelling is based on structural mechanics. Proposed method is proven and effective, but has some disadvantages when it is used in nanocomposite material FEM modelling. In addition, different defects within CNTs structure, as well as the waviness of CNTs, greatly influence the mechanical properties of CNTs. The paper at hand proposes the equivalent beam model for modelling the single walled carbon nanotubes (SWNT), straight and waved, with vacancy defects within structure.

1. Introduction

A great deal of scientific papers, with both theoretical and experimental research on single and multi walled carbon nanotubes (SWNT; MWNT), has been published [1, 2, 3]. Common conclusion of all the researches is that CNTs possess extraordinary material properties, which makes them an excellent choice as a composite material reinforcement [4, 5]. Their nano size is a source of a lot of problems when it comes to investigating CNTs mechanical properties, so numerical methods and computer assisted modelling emerges as logical solutions for given quest. One of the tested and proven numerical method is based on structural mechanics. In proposed method, since the CNT is basically a space frame structure, or molecule, consisting from carbon atoms and covalent bonds, the covalent bonds are replaced with the beam finite elements, and atoms with nodes. [6, 7, 8]. That gives approximately 2350 finite elements per one SWNT (with the length around 13,77 nm), depending on CNT pattern. The problem arises when this CNT model is to be used in nanocomposite modelling where, in addition to CNT finite model, one has to model matrix of nanocomposite and matrix – reinforcement (CNT) interaction. The number of finite elements in such nanocomposite model arises drastically, due to interaction modelling by using nonlinear rod elements [8, 9]. Multiscale modelling methods are smooth way to overcome this issue, but nevertheless, the proposed equivalent beam model of CNT can also be an effective solution. Early research on CNTs was directed toward determination of mechanical properties of ideally shaped CNTs [10, 11, 12], but more exact research detected that SWNT and MWNT carbon nanotubes rarely come in ideal form and without defects. Regardless of the manner of production control and methods, CNTs are usually waved [13, 14], and studies have shown that the waviness of the nanotube has an influence on the final mechanical properties of CNTs, namely longitudinal elastic modulus, and, thus, nanocomposite materials [15, 16, 17]. Another common problem in nanotube structure is appearance of various defects [18, 19, 20], which can occur either naturally or artificially. The impact of mentioned imperfections on longitudinal elastic moduli of SWNT and DWNT was investigated in authors' previous work [21], and those obtained results are basis for determination of elastic moduli for proposed equivalent SWNT beam model, presented in this paper. The armchair (5, 5) and zig-zag (9, 0) pattern of straight and waved SWNTs, with different vacancy defects are considered.

2. SWNT examples

Carbon nanotubes have very high aspect ratio (diameter/length ratio), with diameter in order of nanometers and length in micrometers. High aspect ratio leads to waved shape of CNTs, which is experimentally proved [18, 19], and to small bending stiffness. Waviness of CNTs is in literature defined with the waviness ratio w [22], as ratio between wave amplitude and wavelength. However, mechanical properties of CNTs are affected also by various defects within nanotube structure, which can occur either naturally or artificially. The incomplete bonding defects, i.e. vacancies or missing atoms in CNT structure, are influential when it comes to final mechanical properties of carbon nanotubes, especially longitudinal elastic modulus [21]. Table 1. shows the degradation of longitudinal elastic modulus value, with the increase of the both, the waviness ratio w and the number of vacancies within CNT structure, for both patterns; armchair and zig-zag. These results are obtained from authors' previous work [21]. The CNT models used in aforementioned paper are briefly described as follows. Armchair (5, 5) and zig-zag (9, 0) patterns are selected for SWNT, with similar diameter (0,678nm – 0,695nm) and aspect ratio (20,2). All CNTs are modelled as a space frame structure (atoms replaced by nodes, covalent bonds with the beam finite elements [6, 7, 8]). Using eigenvalue analysis, the waved shape of the nanotubes were obtained. Four SWNT models were prepared and used, straight and three waved, with waviness ratios 0.03; 0.05 and 0.08, as shown in figure 1.

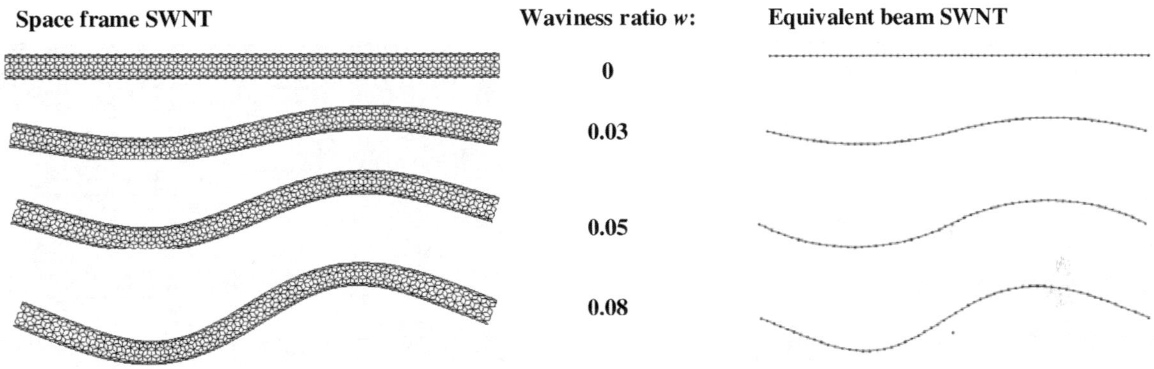

a) b)

Figure 1. Examples of used SWNT models.

Vacancies in SWNTs are introduced by removing one node and corresponding three beam elements. Respectively, the given percentage of missing atoms (0,1%, 0,5%, 1%, 2%, 5%, 10%), randomly selected, were removed from SWNTs structure. To determine the elastic modulus, an axial loading case was used, with 1 nN axial force used in all examples. The longitudinal elastic modulus was obtained using classical term $E = (F \cdot l / A \cdot \Delta l)$, where l represents nanotube length and Δl the elongation obtained from FEM analysis. The obtained results are shown in the table 1. The results for straight SWNT are the basis for modelling the SWNTs using equivalent beam approach. SWNT models with same waviness ratio were prepared, but using only beam finite elements which represent SWNT geometry (Fig 1b.). The beam finite element property was adjusted and prepared in the manner that they coincide with carbon nanotube properties, i.e. cross sectional area and moment of inertia. The data obtained using space frame approach, for straight SWNT, namely the elastic modulus values, were used as relevant in defining the material characteristics of beam finite element property. The number of finite elements per one SWNT drops to only 57 elements, as opposed to 2350 finite

IOP Conf. Series: Materials Science and Engineering **625** (2019) 012004 doi:10.1088/1757-899X/625/1/012004

elements in space frame model of SWNT. After the conducted analysis, same axial loading case was used to obtain the values for the longitudinal elastic moduli of equivalent beam SWNT model, and those results are given in table 2. The analysis was carried out for armchair and zig-zag pattern, with the same waviness ratio and vacancies percentage.

3. Results

Results for longitudinal elastic modulus obtained from aforementioned examples are given in following tables and figures. Tables 1. and 2. show the results for armchair and zig-zag SWNT, space frame approach and equivalent beam approach, respectively. As it is been shown before, there is a noticeable drop of longitudinal elastic modulus with the increase of the waviness ratio of the nanotube, regardless of the CNT type and pattern. The value of elastic modulus also decreases with the increase of the vacancies within the same waviness ratio of the CNT, but is practically unaffected when percentage of defects is less than 1%. The same behaviour pattern can be observed in equivalent beam approach models, figure 2. and 3.

Table 1. Results for space frame models of SWNT

	Longitudinal elastic modulus E, GPa							
	Waviness ratio, w (armchair):				Waviness ratio, w (zig-zag):			
Vacancy %:	0	0.03	0.05	0.08	0	0.03	0.05	0.08
No defects	1048.76	341.58	135.09	74.27	1030.99	349.11	139.97	77.45
0.1%	1008.82	335.58	132.74	72.96	987.36	342.38	138.77	73.05
0.5%	979.79	316.59	127.16	71.98	918.42	328.33	129.49	74.78
1%	908.78	313.61	126.50	69.89	848.07	310.91	124.27	73.59
2%	796.56	285.31	117.67	60.25	818.34	299.12	97.38	56.66
5%	662.10	219.99	76.94	47.99	446.47	87.64	103.91	47.93
10%	146.87	74.42	59.98	13.91	282.99	60.26	36.85	18.77

Table 2. Results for equivalent beam models of SWNT

	Longitudinal elastic modulus E, GPa							
	Waviness ratio, w (armchair):				Waviness ratio, w (zig-zag):			
Vacancy %:	0	0.03	0.05	0.08	0	0.03	0.05	0.08
No defects	1048.76	375.87	166.57	97.44	1030.98	381.31	169.69	99.22
0.1%	1008.82	361.56	160.23	93.73	987.36	365.18	162.51	95.02
0.5%	979.79	351.16	155.62	91.03	918.42	339.68	151.16	88.39
1%	908.78	325.71	144.18	84.43	848.07	313.66	139.58	81.62
2%	796.56	285.49	126.51	74.01	818.34	302.66	134.69	78.75
5%	662.10	237.29	105.16	61.52	446.47	165.13	73.48	42.97
10%	146.87	52.64	23.33	13.65	282.97	104.66	46.57	27.23

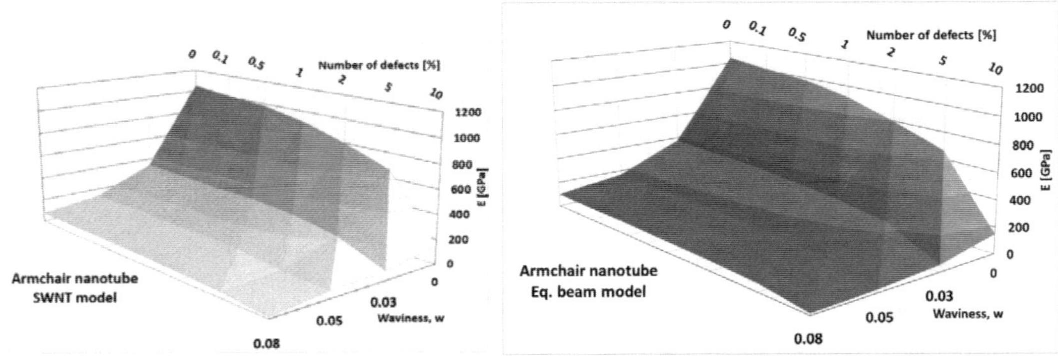

a) b)

Figure 2. Change of: elastic modulus/waviness ratio/number of defects for Armchair SWNT: space frame model (a), equivalent beam model (b).

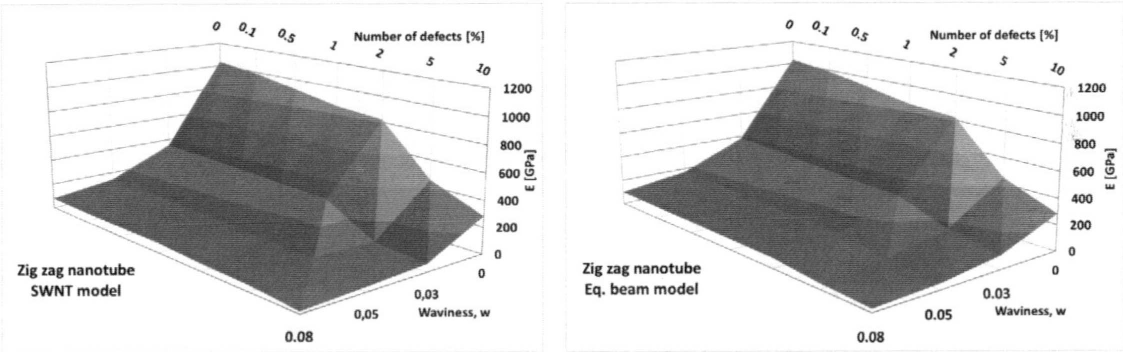

b) b)

Figure 3. Change of: elastic modulus/waviness ratio/number of defects for Zig-zag SWNT: space frame model (a), equivalent beam model (b).

4. Conclusions

The equivalent beam modelling of single walled carbon nanotube approach is shown and analysed in this paper, on finite element model of armchair and zig-zag SWNT pattern. This was done on nanotubes with different waviness ratio and with different number of vacancies within structure. The main goal and idea was to replace the space frame model of SWNT with simpler, equivalent beam model, but to retain all of the characteristics and benefits of space frame model of SWNT, or better said to retain all of the characteristics of the real carbon nanotube. The obtained results confirm the validity of equivalent beam model, but it has to be emphasized that presented models are adequate for small strain problems. Presented equivalent beam model will be useful in future research on theoretical modelling of nanocomposite materials, reinforced with SWNT, since the number of finite elements in the nanocomposite model will be significantly lesser, and thus, the new approaches offer.

Acknowledgments

This work has been supported by the University of Rijeka under the project no. 17.10.2.1.03. This support is gratefully acknowledged.

References

[1] Xie S, Li W, Pan Z, Chang B, Sun L 2000 *J Phys Chem Solids* **61** 1153-1158

IOP Conf. Series: Materials Science and Engineering **625** (2019) 012004 doi:10.1088/1757-899X/625/1/012004

[2] Ruoff RS, Qian D, Liu WK 2003 *Physique* **4** 993-1008
[3] Wan H, Delale F 2010 *Meccanica* **45** 43-51
[4] Tserpes K I, Papanikos P, Labeas G, Pantelakis S G 2008 *Theoretical and Applied Fracture Mechanics* **49** 51-60
[5] Murarescu M, Dima D, Andrei G, Ciriumaru A 2011 *Annals of DAAAM for 2011 & Proceedings of 22nd International DAAAM Symposium* **22** 925-6
[6] Brcic M, Canadija M, Brnic J, Lanc D, Krscanski S, Vukelic G 2009 *Estonian Journal of Engineering* **15** 77-86
[7] Canadija M, Brcic M, Brnic J 2013 *Review on Advanced Materials Science* **33** 1-6
[8] Brcic M, Canadija M, Brnic J 2013 *Meccanica* **48** 2209-20
[9] Canadija M, Brcic M, Brnic J 2017 *Meccanica* **53** 1655-1668
[10] Fisher F T, Bradshaw R D, Brinson L C 2002 *Applied Physics Letters* **80** 4647-49
[11] Fisher F T, Bradshaw R D, Brinson L C 2003 *Composites Science and Technology* **63** 1689-1703
[12] Bradshaw R D, Fisher F T, Brinson L C 2003 *Composites Science and Technology* **63** 1705-22
[13] Ebbesen T, Takada T 1995 *Carbon* **33** 973-8
[14] Shokrieh M M, Rafiee R 2010 *Composite Structure* **92** 647-52
[15] Jiang L Y, Huang Y, Jiang H, Ravichandran G, Gao H, Hwang K C 2006 *Journal of the Mechanics and Physics of Solids* **54** 2436-52
[16] Tan H, Jiang L Y, Huang Y, Liu B, Hwang K C 2007 *Composite Science and Technology* **67** 2941-46
[17] Qian D, Wagner G J, Liu W K, Yu M, Ruoff R S 2002 *Appl. Mech. Rev* **55** 495-533
[18] Shaffer M S P, Windle A H 1999 *Adv. Mater.* **11** 937-41
[19] Vigolo B, Penicaud A, Coulon C, Sauder C, Pailler R, Journet C, Bernier P, Poulin P 2000 *Science* **290** 1331-34
[20] Patel A M, Joshi A Y 2016 *Procedia Technology* **23** 122-9
[21] Brcic M, Canadija M, Brnic J 2018 *IOP Conf. Series: Materials Science and Engineering* **378**
[22] Farsadi M, Ochsner A, Rahmandoust M 2012 *Journal of Composite Materials* **47** 1-10

IOP Conf. Series: Materials Science and Engineering **625** (2019) 012005 doi:10.1088/1757-899X/625/1/012005

Feasibility of producing unfired four-hole hollow bricks from blended cement-fly ash-chippings under coupled-static forming pressure

Trong-Phuoc Huynh[1], Ngoc-Duy Do[2], Chao-Lung Hwang[3], Si-Huy Ngo[4] and Ngoc-Hang Thi Mai[4]

[1] Department of Rural Technology, College of Rural Development, Can Tho University, Campus II, 3/2 St., Ninh Kieu Dist., Can Tho City 900000, Vietnam

[2] Department of Civil Engineering, College of Engineering Technology, Can Tho University, Campus II, 3/2 Street, Ninh Kieu District, Can Tho City 900000, Vietnam

[3] Department of Civil and Construction Engineering, National Taiwan University of Science and Technology, No. 43, Keelung Rd., Da'an Dist., Taipei City 10607, Taiwan

[4] Department of Engineering and Technology, Hong Duc University, No. 565, Quang Trung Street, Dong Ve Ward, Thanh Hoa City 440000, Vietnam

E-mail: mikehwang@mail.ntust.edu.tw

Abstract. Unfired four-hole hollow brick (UFHB) is a type of environmental-friendly brick with low energy consumption. However, the using of ordinary Portland cement (OPC) in UFHB is costly and indirectly affects the environment. This paper focuses on an exploratory study on the use of blended fly ash (FA)-OPC in the production of the UFHB and proposes the application of FA-OPC mixtures as a timely solution for the reduction of OPC in UFHB. The 80×80×180 mm UFHB samples were produced using different proportions of FA and OPC under a constant coupled-static forming pressure. The report exhibited experimental characterizations on physical and mechanical properties of both starting materials and UFHB products. In addition, a scanning electron microscope (SEM) analysis was used to evaluate the microstructure of the final UFHB. Test results show that the inclusion of FA in the UFHB mixtures provided positive effects on brick properties. As a result, 10% FA was found as an optimal content, which resulted in the highest brick's strength, the lowest water absorption rate, and a denser microstructure. Furthermore, properties of all of the UFHB samples produced for this study satisfied the requirements of the National Vietnamese standard for non-bearing building brick.

1. Introduction

Vietnam is currently on the verge of large-scale urbanization hence facing huge demand for building houses and thus for the materials to build them. To cope up with this housing necessity, the country needs to use a million units of bricks per day. It is continually expanding on account of a rapid increase in demand for bricks in infrastructure and housing industry. To produce a traditional fired clay brick, a large number of natural sources and energy consumption are used. Therefore, the invention of unfired brick (UB) is necessary in order to limit the use of natural materials, as well as reduce the harmful effects on the environment. However, the production of UB spent a large amount of ordinary Portland cement (OPC) and it indirectly affects the environment as well [1, 2]. This underlines the need to find an eco-friendly substitute to OPC in order to develop an alternative green material. In recent decades, the application of fly ash (FA), a by-product of high-temperature combustion of coal in coal-fired power plants, has increased many folds [3, 4]. FA is not only a pozzolanic material (chemical effect) but also a micro-filler (physical effect) that greatly contributes to enhancing the properties of the final products [5, 6]. So far, the utilization of FA in the production of UB and the study on characteristics of the UB have still been limited. Therefore, this study investigates the feasibility of producing a new type of UB, so-called unfired four-hole hollow bricks (UFHB), using a mixture of OPC, FA, and chippings. Especially, a coupled-static forming pressure was applied to form the UFHB samples. The effects various compositions of OPC and FA on engineering properties and microstructure of the UFHB samples were also studied.

2. Experimental works

2.1. Material properties

Locally grade 40-OPC and type F-FA were used as binder materials. Characteristics of both FA and OPC are shown in Table 1. It can be seen that the major mineralogical compositions of OPC were SiO_2 and CaO, whereas the major compositions of FA were SiO_2 and Al_2O_3, which may react with other products of cement's hydration under $Ca(OH)_2$ environment to form C-S-H gel [7].

Table 1. Characteristics of binder materials.

Items		OPC	FA
Specific gravity		3.15	2.29
Mean particle size (μm)		19.1	21.5
Specific surface area (m^2/g)		0.78	0.66
Strength activity index at 28 days (SAI, %)		100	86.5
Chemical compositions (wt.%)	SiO_2	20.04	64.01
	Al_2O_3	4.24	22.14
	Fe_2O_3	3.12	5.64
	CaO	62.43	2.75
	MgO	4.17	0.92
	SO_3	2.97	0.61
	K_2O	0.43	1.36
	Na_2O	0.33	0.85

In addition, it can be clearly observed that the particle size of FA was almost similar to that of OPC (Table 1) and the FA had morphology form of spherical shape with different particle sizes (Figure 1). As usual, the smaller particle size results in a higher degree of reaction, indicating by a high SAI value of FA (86.5%) compared to OPC (Table 1).

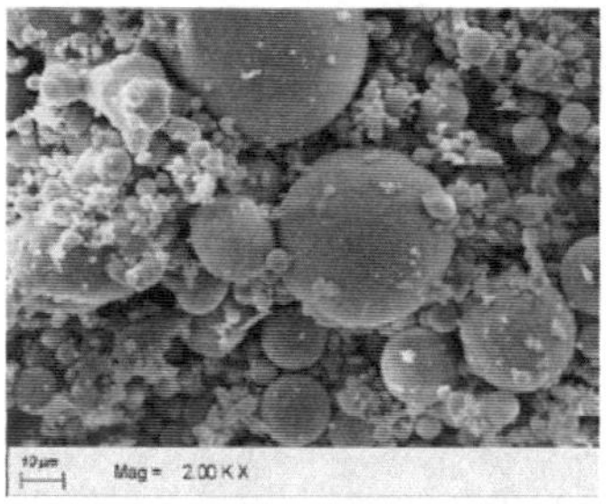

Figure 1. SEM image of FA particles.

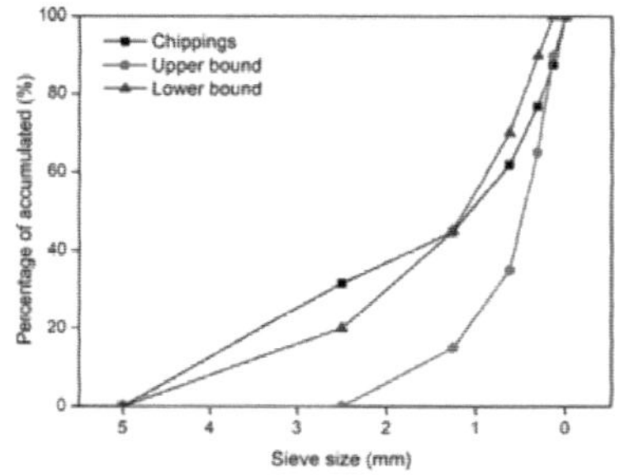

Figure 2. Gradation curve of chippings.

Figure 3. The appearance of UFHB samples.

Chipping, in air-dry condition, with a maximum particle size of 5 mm, a density of 2550 kg/m^3, and water absorption rate of 3.1% was adopted as fine aggregate to produce the UFHB samples. The gradation curve of chippings is plotted in Figure 2. It is noted that the chipping with a particle size of 2.5 mm that accounted for more than 30% will form the resistance structure and contribute to enhancing the samples'strength. Furthermore, the appearance of the smaller particle size of below 0.63 mm will play as a filler and thus improve the physical properties of the brick.

2.2. Mix proportioning

In the present study, a total matrix of 6 UFHB mixtures was prepared. A fixed water-to-binder (w/b) ratio of 0.2 was used for all of the UFHB mixtures. The UFHB mixtures were proportioned to study the effects of different OPC and FA contents on the brick's properties. The first three UFHB mixtures were prepared with various OPC contents of 8, 10, and 12%, while the rest three UFHB mixtures were prepared with the addition of FA to the mixtures at 10%, 15%, and 20%. The mix proportions for the preparation of the UFHB samples are shown in Table 2.

Table 2. Ingredient proportions for making samples.

Materials (kg/m^3)	Brick mixtures					
	C08F00	C10F00	C12F00	C10F10	C10F15	C10F20
OPC	195.4	242.9	289.8	229.7	223.6	217.8
FA	-	-	-	229.7	335.4	435.7
Chipping	2247.2	2185.8	2125.1	1837.3	1676.9	1524.8
Water	39.1	48.6	58.0	91.9	111.8	130.7

Note: *C represents the OPC content and F represents the FA content.*

2.3. Samples preparation and test methods

All of the raw materials were prepared in accordance with Table 2. The dry materials were uniformly mixed using a laboratory mixer. The mixture was then filled up in a steel mold and was compacted under a coupled-static forming pressure of about 4.7 MPa to form the UFHB samples as presented in Figure 3. The brick samples were de-molded and placed in air for 24 h before fully immersed in water for 30 min and then covered by a thin plastic layer to keep the moisture until the testing time. The 28-day UFHB samples were subjected to the tests of compressive strength and water absorption following the guidelines of the National Vietnamese standard [8]. Furthermore, the broken specimens from the compression test were collected to evaluating the microstructure of the brick using SEM technique.

3. Results and discussion

3.1. Compressive strength

The 28-day compressive strength values of the UFHB samples with different OPC and FA contents are presented in Figure 4. The compressive strength increased proportionally OPC content. The brick's strength of 4.24 MPa was obtained at 8% OPC and the compressive strength value increased by 4.0% and 9.4% with increasing OPC content to 10% and 12%, respectively. Moreover, it is expected that the inclusion of FA improved the compressive strength of the UFHB samples. The bricks with 10%, 15%, and 20% FA had compressive strength value of 36.7%, 17.7%, and 3.4% higher than the free-FA bricks. The fact that FA is a pozzolanic material with high SiO_2 content and high SAI (Table 1), which is the strength determinant in combination with CaO in OPC, forming hydrated lime ($Ca(OH)_2$) in the presence of water and resulting in the formation of calcium silicate hydrate (C-S-H) gel. This gel is responsible for the strength gain of the samples. Moreover, some small unreacted FA particles will play as a filler material to reduce the void between larger particles in structure. Hence, the presence of FA in the mixture increased the compressive strength of the UFHB. However, an excess amount of FA caused a reduction in brick's strength. It may due to the presence of more unreacted FA particles, which resulted in a low degree of reaction and thus less formation of C-S-H gel in the matrix. The UFHB samples produced in this study satisfied for non-bearing wall construction as stipulated by TCVN 6477-2016 [8].

3.2. Water absorption

Figure 5 presents the water absorption (WA) levels of the UFHB samples. Obviously, the WA decreased with increasing OPC content. The inclusion of FA in brick mixtures was also attributable to the low WA rate of the bricks. However, the WA levels increased with FA content. This finding was supported by Bigas and Gallias [9]. It can be clearly perceived that more OPC content will generate more bonding agent (C-S-H gel) to create a denser structure and thus reduces the WA rate of the system. As the result, all of the UFHB samples produced in this study had the WA rates of below maximum limit of 12% as required by TCVN 6477-2016 [8]. Further, it is interesting to state that compressive strength was inversely proportional to the WA (Figure 6). It is attributed to the fact that the higher the compressive strength, the lower the porosity and WA of the brick was. The correlation between compressive strength and WA of the bricks was presented by the equation of $y = 155.151 - 33.888x + 1.902x^2$ with R^2 of 0.93. As the previous discussion, the inclusion of FA reduced the void within the brick structure and consequently reduced the WA rate of the bricks. However, more voids/ pores were introduced with the presence of access amount of FA (Figure 7), which resulted in a high level of WA.

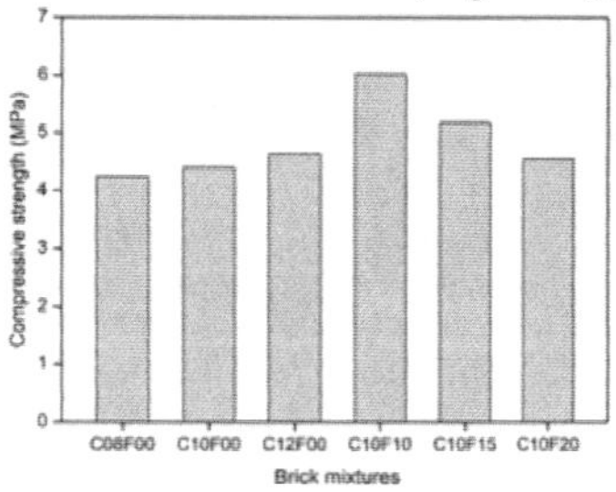

Figure 4. Compressive strength of brick samples at 28 days.

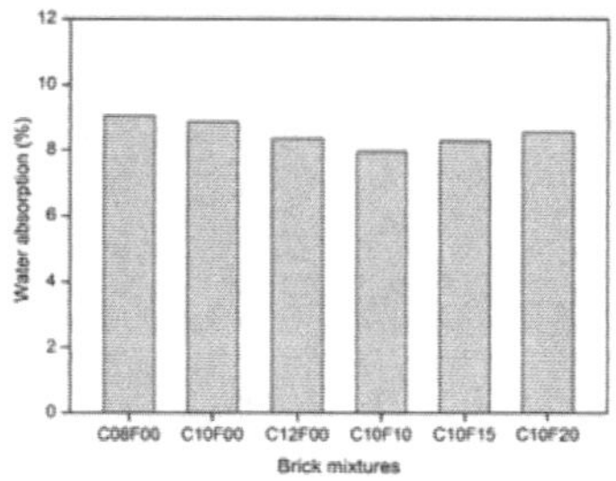

Figure 5. Water absorption of brick samples at 28 days.

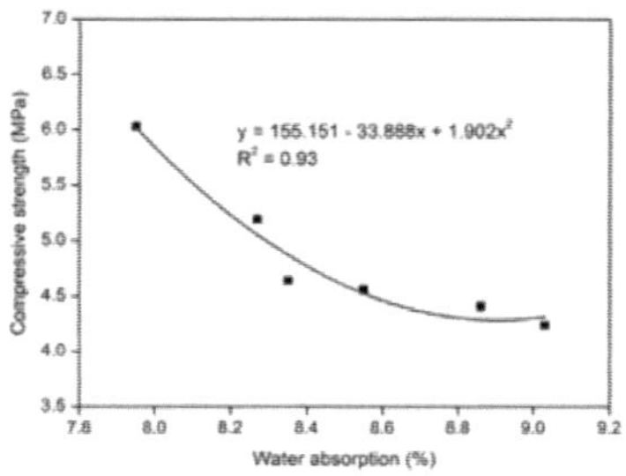

Figure 6. The relationship between WA and compressive strength of bricks.

3.3. SEM observation

The SEM micrographs of the 28-day-old UFHB specimens are shown in Figure 7. This figure provided information on the degree of reaction within the system, indicated by the change in the material particle shape, the rearrangement of the system, and the presence of partial/ unreacted particles.

(a) C08F00

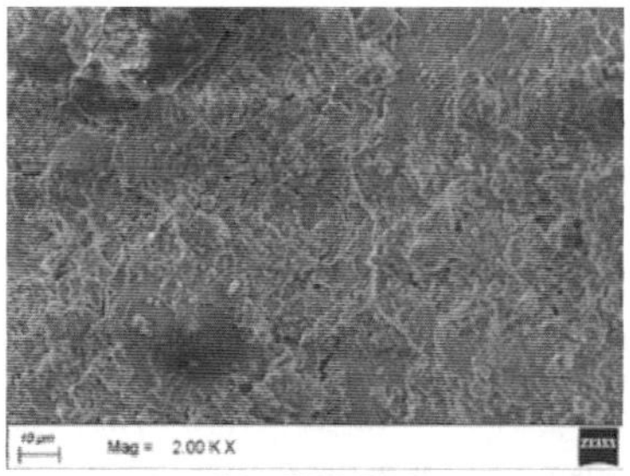

(b) C10F00

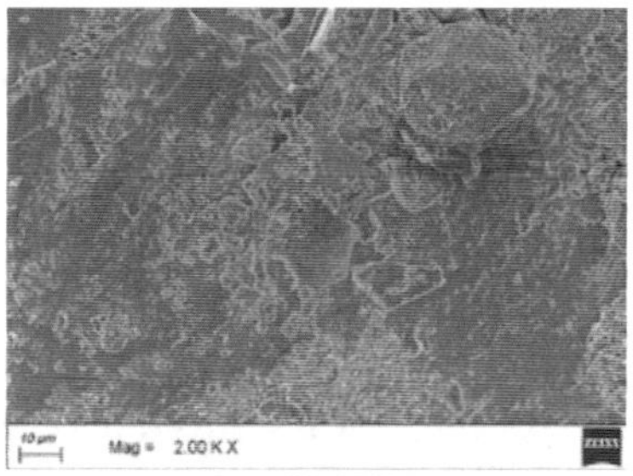

(c) C12F00

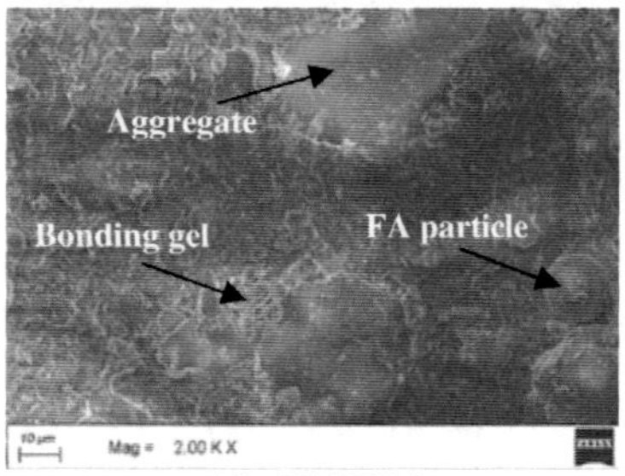

(d) C10F10

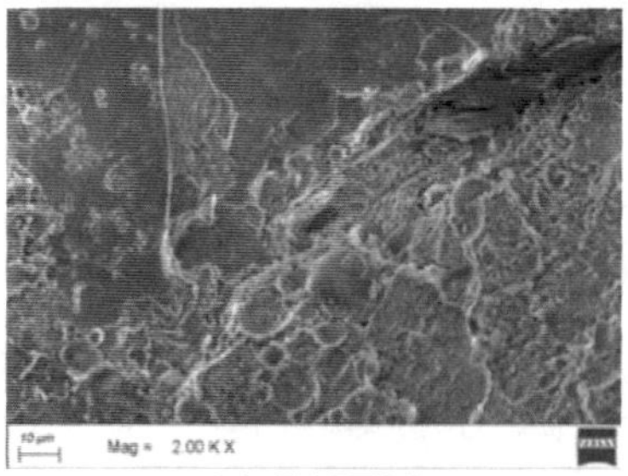

(e) C10F15

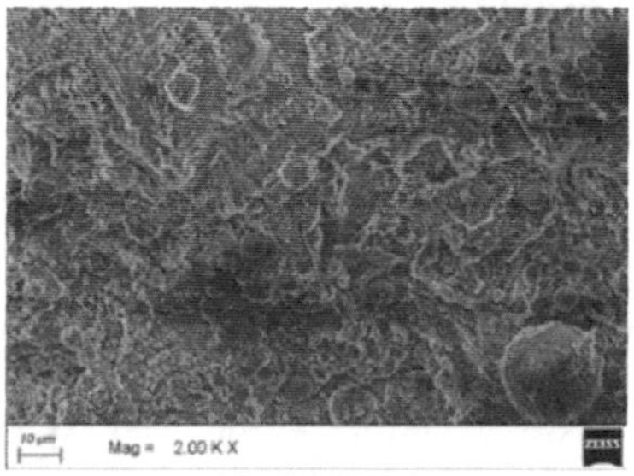

(f) C10F20

Figure 7. SEM images of brick samples.

There were three distinguishable phases that were detected from the micrographs, including aggregate particles (smooth grey areas), unreacted FA particles (round grey dots), and bonding gels between the particles (textured light grey areas). It can be clearly seen that increasing the OPC content from 8% to 10% and 12% resulted in an increased quantity of bonding gel as the result of cement hydration. This contributed to the higher strength of the UFHB samples as previously presented in Figure 4. On the other hand, the structure of the 10% FA brick was much denser than that of the others. Increasing the FA content to 15% and 20% caused a less dense structure with many unreacted particles and pores, leading to the lower strength and higher WA rate of the bricks. This observation was in good agreement with the results of the compressive strength and WA of the UFHB samples as aforementioned.

4. Conclusions

The present experimental works investigated the feasibility of producing UFHB using various proportions of OPC, FA, and chippings. Based on the obtained results, the following conclusions can be drawn:

1. The compressive strength of the UFHB samples increased proportionally with OPC content, whereas the reverse trend was observed for WA rates of the bricks.
2. The inclusion of FA in the brick mixtures provided a positive effect of on the properties of the UFHB samples. All of the FA-bricks had higher strength values and lower WA levels as compared to the free FA-brick and the respective maximum strength value and the lowest WA rate of 6.03 MPa and 7.95% were achieved at the brick containing 10% FA.
3. SEM observation found that the UFHB samples with 10% FA content exhibited a denser structure relative to the other samples. This finding agreed with the compressive strength of the bricks.
4. The results of this study encourage the use of raw FA in the manufacture of unfired building bricks and particularly demonstrated great feasibility of producing UFHB with the utilization of FA, OPC, and chippings. The use of such materials is not only cost effective but also reduces the negative impact on the environment.
5. Further research may be conducted using other types of industrial solid waste materials in combination with OPC to produce UFHB in either laboratory scale or industrial scale.

Acknowledgments
The authors gratefully acknowledge the valuable assistance provided by Trung Hau Manufacturing Co. Ltd. (Vietnam) in the preparation of the UFHB samples used for this study.

References

[1] Li H, Xu W, Yang X, and Wu J Preparation of Portland cement with sugar filter mud as lime-based raw material 2014 *J. Clean. Prod.* **66** 107–112
[2] Mikulčić H, Klemeš J J, Vujanović M, Urbaniec K, and Duić N Reducing greenhouse gases emissions by fostering the deployment of alternative raw materials and energy sources in the cleaner cement manufacturing process 2016 *J. Clean. Prod.* **136** 119–132
[3] Ahmaruzzaman M A review on the utilization of fly ash 2010 *Prog. Energy Combust. Sci.* **36** 327–363
[4] Shemi A, Mpana R N, Ndlovu S, van Dyk L D, Sibanda V, and Seepe L Alternative techniques for extracting alumina from coal fly ash 2012 *Miner. Eng.* **34** 30–37
[5] Poon C S, Lam L, and Wong Y L A study on high strength concrete prepared with large volumes of low calcium fly ash 2000 *Cem. Concr. Res.* **30** 447–455
[6] Wang A, Zhang A, and Sun W Fly ash effects: III. The micro-aggregate effect of fly ash 2004 *Cem. Concr. Res.* **34** 2061–2066
[7] Hewlett P C 2003 *Lea's chemistry of cement and concrete* Elsevier
[8] TCVN 6477-2016 2016 *Concrete bricks* Ministry of Construction (Vietnam)
[9] Bigas J P and Gallias J L Effect of fine mineral additions on granular packing of cement mixtures 2002 *Mag. Concr. Res.* **54** 155–164

Comparison of human and bovine dental enamel by TEM and t-EBSD investigations

A. Koblischka-Veneva[1,2], M. R. Koblischka[1,2], J. Schmauch[1], M. Hannig[3]

[1] Experimental Physics, Saarland University, P.O.Box 151150, 66041 Saarbrücken, Germany
[2] Superconducting Materials Laboratory, Department of Materials Science and Engineering, Shibaura Institute of Technology, Tokyo 135-8548, Japan
[3] Clinic of Operative Dentistry, Periodontology and Preventive Dentistry, Saarland University, Building 73, 66421 Homburg/Saar, Germany

E-mail: anjela@shibaura-it.ac.jp

Abstract. The microstructures of human dental enamel and bovine enamel are compared to each other. To obtain samples for transmission electron microscopy (TEM), focused ion-beam (FIB) milling is used. The preparation of such TEM-slices is found to be very effective when operating the FIB with adapted parameters. After the milling process, the TEM-slices are then thinned by means of the ion beam to achieve samples being transparent for the electron beam. With a home-built sample holder, the electron backscatter diffraction (EBSD) can be operated in the scanning electron microscope (SEM) in transmission mode called t-EBSD. This technique enables the crystallographic orientation measurement on nanometer-sized, non-conducting enamel grains with a reasonable quality. Both TEM and t-EBSD images reveal a similar arrangement of the apatite crystals within the enamel, but it is obvious that the nanostructure of human enamel follows a more complex construction principle. The grain sizes of bovine enamel are much larger, and it is difficult to recognize the chain arrangement as found previously in the human enamel. As a result of the comparison, one can state that the nanostructure of human enamel is clearly more complicated than the bovine counterpart.

1. Introduction

The hierarchial micro- and nanostructure of tooth enamel can be considered as a masterpiece of natural nanotechnology [1]. The tooth enamel is an important natural biomaterial with unique properties, including hardness and life-long durability [2–4]. The enamel is built from calcium hydroxyapatite, $[Ca_{10}(PO_4)_6(OH)_2]$ (HAP). On the microscale, the enamel containis highly organized prisms, forming a pillar-like structure. While this structural arrangement was investigated with many different imaging techniques, there is only little knowledge about the HAP arrangement on the nanoscale. Recently, good quality TEM images could be obtained from TEM slices prepared by focused ion-beam milling [5]. Moreover, these TEM slices could be employed for transmission electron backscatter diffraction (t-EBSD) investigations enabling the measurement of HAP grain orientations with a resolution in the 20 nm range. From these observations, it became possible to deduce important principles of how the nanostructure is built up.

The electron backscatter diffraction (EBSD) technique has proven to be a very useful tool mostly on metallic samples, and more recently, also on various ceramic sample types

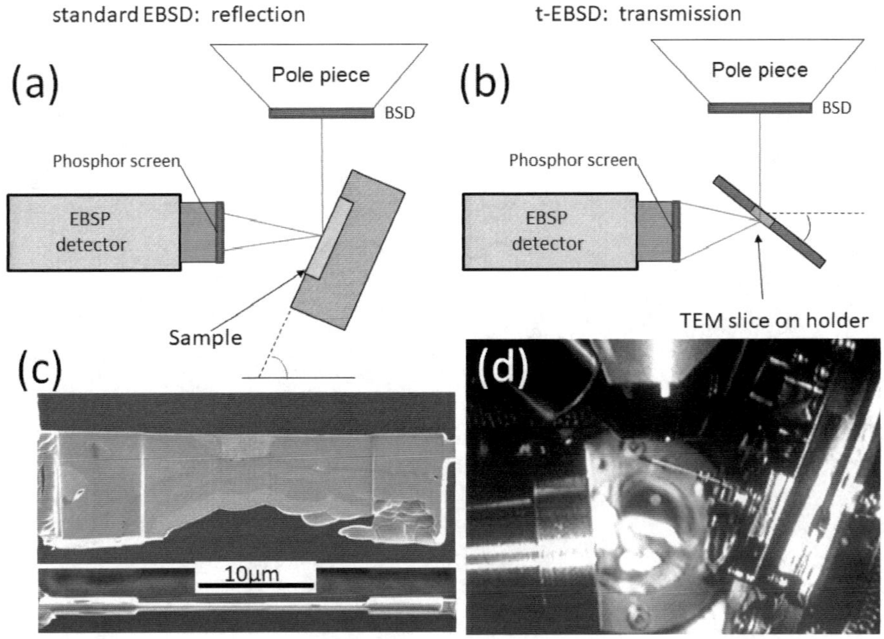

Figure 1. Schematic drawings of conventional EBSD (a) and t-EBSD (b). Also indicated are the sample inclination angles required for EBSD imaging. (c) presents SEM views of the ready TEM slice fixed on the micromanipulator – the surface for analysis (top) and the cross-section (bottom). Finally, (d) shows the home-built TEM slice holder for t-EBSD imaging mounted in the SEM chamber.

[6–9]. However, the standard EBSD technique suffers from distinct problems concerning the observation of nanometer-sized grains. Therefore, the t-EBSD technique was very recently developed by several authors [10–13] with the goal to improve the spatial resolution of EBSD on such materials. As result, the spatial resolution could be improved from several tens of nanometers to ∼5-10 nm, depending on the material to be studied [14, 15]. Furthermore, the effects of charging in non-conducting samples are considerably reduced, which is a great advantage for ceramic or even biomaterials [5]. All these improvements make t-EBSD the method of choice to analyze the crystallographic orientations and grain boundary misorientations of tooth enamel.

The processes in nature are very complex, so it is only understandable that the microstructure of tooth enamel differs for different species. In this contribution, a comparison of human tooth enamel with bovine tooth enamel is performed using TEM and t-EBSD. The bovine enamel is often used in the laboratory as a model system for the human one.

2. Experimental procedures

2.1. Electron microscopy

Figure 1 gives a schematic drawing of the arrangements for conventional EBSD (= reflection mode) and t-EBSD (transmission mode). The primary electron beam hits the sample placed with an inclination towards the beam. After entering the sample, an electron cone is formed due to Bragg reflection at the crystal layers [16]. This electron cone is intercepted by a phosphor screen, where the Kikuchi patterns are recorded. Transmission EBSD works with a transmitted electron beam, and the cone is formed on the backside of the sample. Therefore, an electron-

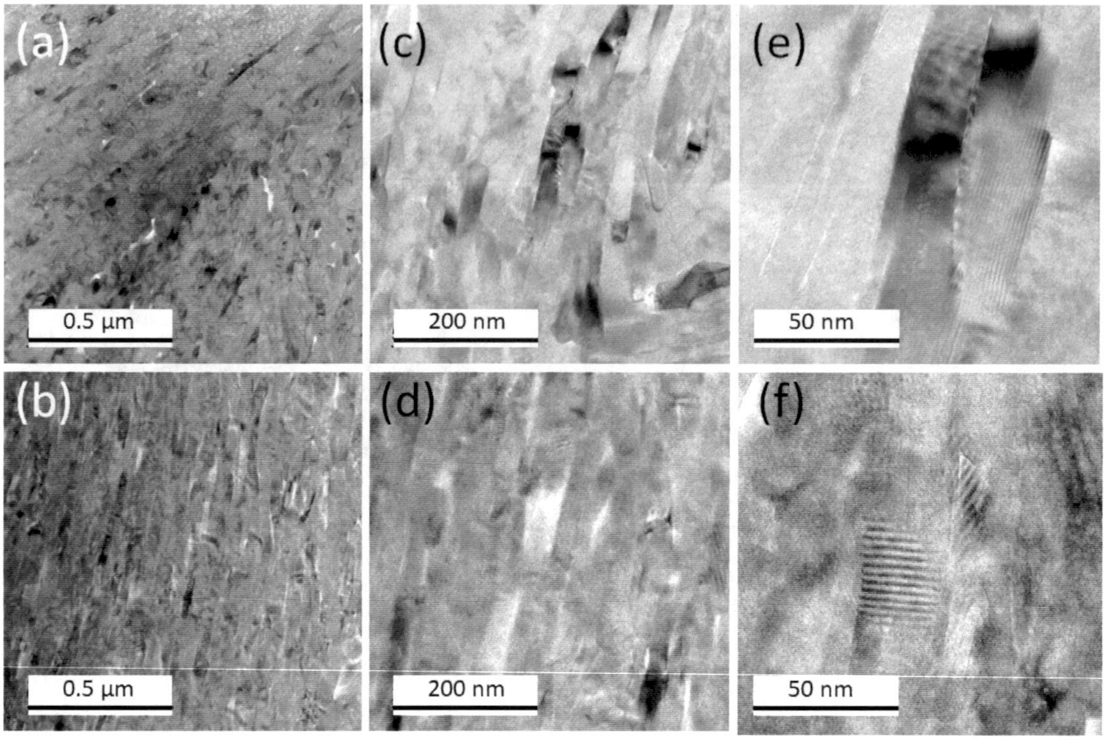

Figure 2. (a), (c), (e): TEM images of the human enamel sample in different magnifications 10000× (a), 25000× (c) and 80000× (d), sample HS2 from Ref. [5]. (b), (d), (f): TEM images of the bovine enamel sample at the same magnifications (sample S8) for comparison.

transparent sample is required for t-EBSD.

TEM investigations were performed using a JEOL JSM-2011 transmission electron microscope operating at 200 KV with a LaB_6 cathode. The EBSD analysis was performed in a JEOL 7000F SEM microscope equipped with a TSL (TexSEM Labs, UT [17]) analysis unit. The Kikuchi patterns were generated at an acceleration voltage of 15 KV, and were recorded by means of a DigiView camera system. To produce a crystallographic orientation map, the electron beam was scanned over a selected surface area and the resulting Kikuchi patterns were indexed and analyzed automatically, represents the common EBSD method working in reflection mode. Automated EBSD scans were performed with a step size down to 50 nm. Although also traditional methods of producing samples for TEM investigations could be used also for t-EBSD, the FIB milling of TEM slices is the method of choice as it enables a proper selection of the area for analysis. Furthermore, the subsequent treatment of the prepared slice with the Ga-ion beam allows larger areas to be analyzed by t-EBSD as compared to the traditional TEM sample preparation methods. For t-EBSD, the TEM-slices were mounted in the SEM on a specially fabricated sample holder allowing for the correct 70° inclination of the sample required for EBSD. The stage with the sample holder is inclined to an angle of -20°, which enables together with the sample mounting the same detector position to be used for the EBSD detector as in the standard configuration. Here, the electron beam is passing through the sample (transmission mode) and the electron cone is formed on the backside of the sample. The electron beam operates at 30 KV, and the working distance is set to 5 mm. The t-EBSD stepsize was chosen to be 5 nm [5,18,19]. An image of the entire arrangement within the SEM chamber is presented in Fig. 1. To improve the imaging quality even further, the TEM slices were treated by additional low-angle argon ion-

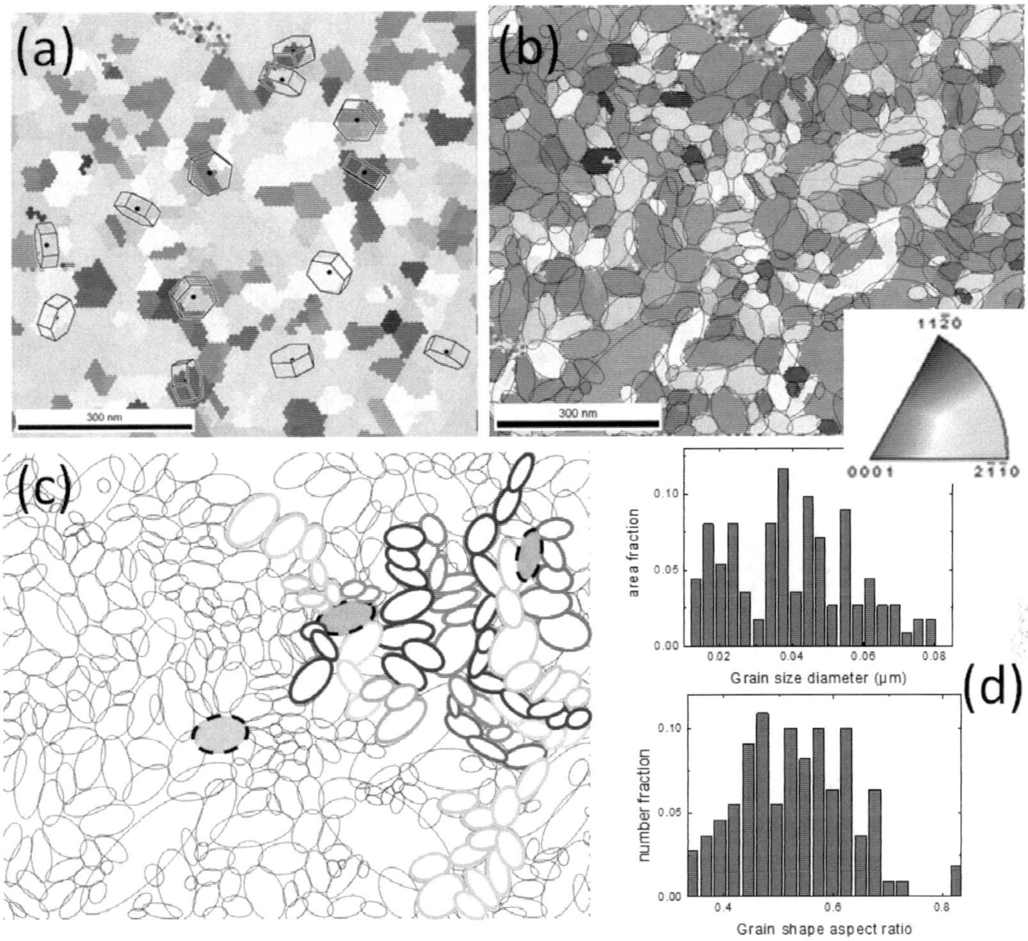

Figure 3. t-EBSD Inverse pole figures (IPF) maps in [001] (a) and [100]-direction (b) of the human enamel sample. The color code for the maps is given in the stereographic triangle. In (a), the orientation of the apatite unit cells is indicated, and in (b) the apatite grains are encircled by ellipses to view the arrangement and the grain shape aspect ratio. (c) represents the central result from Ref. [5], where a building principle of the apatite nanostructure could be deduced by formation of chains of grains with similar orientation starting from nodal points (blue-filled grains). The plots (d) give the EBSD-determined grain size distribution and the distribution of the grain aspect ratio.

polishing (5 KeV, 5 min) to increase the image quality (IQ) of the resulting Kikuchi patterns as described in [20] for the investigation of ferrite samples.

2.2. Surface preparation and TEM slices

For the microstructure analysis, the sample surfaces were polished mechanically using various 3M lapping films (100 μm down to 0.05 μm), and the final treatment was carried out employing Struers SiO_2 (OP-S) suspension with the respective polishing cloth on an automatic polishing machine at a speed of 200 rpm. This procedure yields a total roughness of several nanometers as determined by atomic force microscopy. Further details of the polishing procedure are described in Refs. [5, 9, 21]. No additional carbon layers to enhance the electric conductivity were applied.

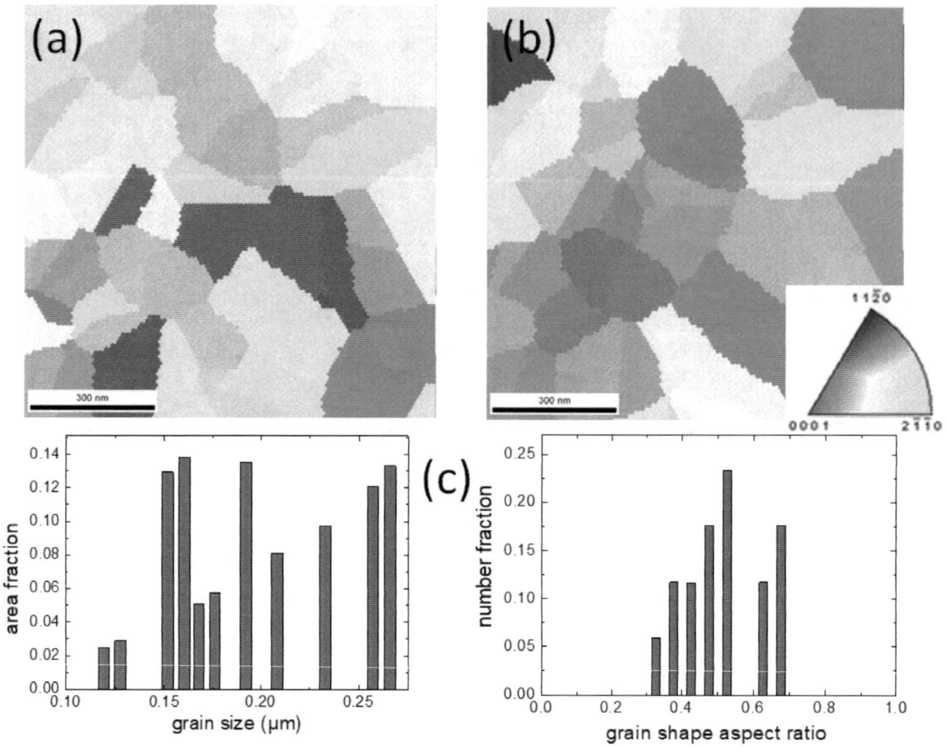

Figure 4. (a,b): IPF maps in [001] and [100]-directions of the bovine enamel sample. The enamel grains are considerably larger as in the human case, but the principal grain orientation is similar. The plots (c) give the EBSD-determined grain size distribution and the distribution of the grain aspect ratio.

The TEM slices for the t-EBSD measurements were produced by focused-ion beam (FIB) milling in a dual-beam FIB workstation (FEI) using a routine allowing for reduced surface damage. The polished sample surface is covered by Pt to reduce the influence of charging effects. After lifting-off the TEM slice from the sample with the micromanipulator, the surface is ion-polished in a separate step by 2 KeV Ga-ions to a thickness of about 80 nm. This step serves to further reduce the preparation damage of the surface area and for a further thinning of the sample to be transparent to the electron beam. Figure 1 (c) presents a typical TEM slice fabricated from a human tooth sample after the completed ion-beam thinning process in a surface view (upper image) and the cross section (lower image). In Fig. 1 (d), the TEM slice is shown ready for t-EBSD investigations in the SEM chamber mounted on the sample holder for t-EBSD.

Two samples were chosen for the present study: The human enamel sample HS2 (Ref. [5]), and the bovine enamel sample S8. Both samples were cut perpendicular to the original tooth surface in an identical way to allow for direct comparison.

3. Results and discussion

Figure 2 presents TEM images at various magnifications (10000×, 25000× and 80000×) from human and bovine enamel. For comparison, the same magnification steps were used in (a), (c), (e) and in (b), (d), (f). The TEM images reveal the dense packing of the HAP grains with no voids between them. All grain boundaries are well defined and clearly resolved in the images. It is further visible that the HAP grains have commonly a long extension. The comparison directly reveals that the bovine HAP grains are larger than the ones of the human enamel sample.

In Fig. 3, the results of the t-EBSD orientation mapping on the human tooth enamel sample HS2 in [001]-direction (a) and [100]-direction (b) are shown. The color code for the HAP orientations is given in the stereographic triangle. The maps reveal the polycrystalline structure of the enamel with many high-angle grain boundaries. The HAP grains are, however, not fully randomly oriented as green and violet orientations are dominating in (a), and the orientation in [001]-direction (red) in (b). Figure 3 (c) represents the main result of Ref. [5]: From the EBSD orientation mappings in [100]-direction, it was deduced that HAP grains of similar orientation form chains which begin at some nodal points from which several chains branch away. This arrangement may provide the necessary strength of the entire nanostructure. As a result, the human enamel structure is a pretty complex one. Finally, Fig. 3 (d) gives the EBSD-determined grain size distribution and the distribution of the grain shape aspect ratio. To determine the latter one, the HAP grains are encircled by ellipses with a major (a) and minor (b) axis. The grain shape aspect ratio is then defined as $g_{\mathrm{ar}} = b/a$, which ranges between 0 and 1 [17].

Figures 4 (a) and (b) give the t-EBSD mapping of the orientations in [001]- and [100]-directions of the bovine enamel sample S8. The color code for the HAP orientations is given in the stereographic triangle. The orientation mapping reveals the polycrystalline grain arrangement with many high-angle grain boundaries, similar to the human enamel. Also here the HAP grains are not oriented fully randomly, even though due the larger grain size the present mapping contains only a small number of HAP grains. Furthermore, on the bovine sample S8 it was not possible to observe a chain formation like in the human enamel sample, nor some nodal points. Figure 4 (c) presents the plots of the grain size distribution and the distribution of the grain shape aspect ratio of the bovine enamel.

The comparison of the plots of Fig. 3 (d) and 4 (c) enable to draw more interesting conclusions: The bovine enamel sample exhibits only a narrow range of the grain shape aspect ratio with values of ~0.5, whereas the human enamel shows a much wider range spanning of g_{ar} between 0.35 and 0.85. This result was obtained on TEM slices of various human tooth samples, so this wider range of g_{ar} may indicate an adaptation of the human teeth for a different use (dietary adaptations omnivore vs. herbivore, [3]).

Using the t-EBSD technique on focused ion-beam prepared TEM slices of enamel stemming from human and bovine teeth, the mapping of the HAP crystallite orientations becomes possible with a spatial resolution in the 10-20 nm range. The measurements clearly reveal the common grain arrangements of the HAP grains in the enamel, but also obvious differences between the two types of teeth are revealed. The grain size, the grain size distribution, and the distribution of the grain shape aspect ratio are different in the two types of teeth. Furthermore, the missing chain formation of the enamel grains in the bovine enamel may be responsible for a different resulting strength of the enamel structure. These observations suggest that the different use of the teeth is reflected in the building scheme of the enamel. This demonstrates that the knowledge about the HAP crystal orientation and shape is essential for the understanding of biomaterials [22].

4. Conclusions

Human enamel is found to be different from bovine enamel (the samples are prepared in the same orientation to enable a comparison). The overall HAP grain orientation is found to be similar for human and bovine enamel in both [001] and [100] orientations. The bovine enamel exhibits only a narrow range of the grain shape aspect ratio at ~0.5, whereas the human enamel shows a much wider range. The grain size of bovine enamel is larger than the human one, and the grain size distribution of the bovine enamel is much more narrow. Furthermore, it was not possible to observe the characteristic chain formation of the enamel grains in the bovine enamel sample. Future work will consider the comparison of the enamel grain orientations of various animals (carnivores, herbivores) with the human enamel, and also the differences between human baby tooth enamel and that of adults.

Acknowledgments

This work was supported by Saarland University ('Anschubfinanzierung'), which is gratefully acknowledged. We also thank F. Soldera (UdS, Institute of Functional Materials) for the excellent FIB work.

References

[1] Hannig M and Hannig C 2010 Nanomaterials in preventive dentistry. *Nat Nanotechnol* **5** 565
[2] Black J, and Hastings G 1998 Handbook of Biomaterial Properties. Springer Verlag Berlin
[3] Macho G A, Jiang Y and Spears I R 2003 Enamel microstructure: A truly three-dimensional structure. *Hum Evol* **45** 81
[4] Cui F Z and Ge J 2007 New observations of the hierarchical structure of human enamel, from nanoscale to microscale. *J Tissue Eng Regenerative Med* **1** 185
[5] Koblischka-Veneva A, Koblischka M R, Schmauch J and Hannig M 2018 Human dental enamel: A natural nanotechnology masterpiece investigated by TEM and t-EBSD. *Nano Research* **11** 3911
[6] Humphreys F J 2004 Characterisation of fine-scale microstructures by electron backscatter diffraction (EBSD). *Scr Mater* **51** 771
[7] Dingley D 2004 Progressive steps in the development of electron backscatter diffraction and orientation imaging microscopy. *J Microsc* **213** 214
[8] Chen D, Kuo J-C and Wu W-T 2011 Effect of microscopic parameters on EBSD spatial resolution. *Ultramicroscopy* **111** 1488
[9] Koblischka M R and Koblischka-Veneva A 2013 Applications of the electron backscatter diffraction technique to ceramic materials *Phase Transit* **86** 651
[10] Trimby P W 2012 Orientation mapping of nanostructured materials using transmission Kikuchi diffraction in the scanning electron microscope. *Ultramicroscopy* **120** 16
[11] Keller R and Geiss R 2012 Transmission EBSD from 10 nm domains in a scanning electron microscope. *J Microsc* **245** 245
[12] Sneddon G C, Trimby P W and Cairney J M 2016 Transmission Kikuchi diffraction in a scanning electron microscope: A review. *Mat Sci Eng R* **110** 1
[13] Suzuki S 2013 Evaluation of transmission-EBSD method and its application to observation of microstructures of metals. *J Jpn Inst Met Mater* **77** 268
[14] Britton B, Holton I, Meaden G and Dingley D 2013 High angular resolution electron backscatter diffraction: Measurement of strain in functional and structural materials. *Microsc Anal* **27** 8
[15] van Bremen R, Ribas Gomes D, de Jeer L T H, Ocelik V and De Hosson J Th M 2016 On the optimum resolution of transmission-electron backscattered diffraction (t-EBSD). *Ultramicroscopy* **160** 256
[16] Reimer L 1985 Scanning Electron Microscopy: Physics of Image Formation and Microanalysis. Springer Science: Berlin Heidelberg
[17] TexSEM Laboratories (TSL). Orientation Imaging Microscopy Software V4.1, User Manual, TexSEM laboratories, (TSL), Draper, UT, 2004.
[18] Koblischka-Veneva A, Koblischka M R, Schmauch J, Inoue K, Muralidhar M, Berger K and Noudem J 2016 EBSD analysis of MgB_2 bulk superconductors. *Supercond Sci Technol* **29** 044007
[19] Koblischka-Veneva A, Koblischka M R, Zeng X L, Schmauch J and Hartmann U 2018 TEM and electron backscatter diraction analysis (EBSD) on superconducting nanowires. *J Phys: Conf Ser* **1054** 012005
[20] Koblischka-Veneva A, Koblischka M R, Schmauch J, Chen Y and Harris V G 2010 EBSD analysis of the microtexture of Ba-hexaferrite samples. *J Phys: Conf Ser* **200** 082014
[21] Koblischka-Veneva A and Koblischka M R 2004 Surface preparation of high-T_c superconductors for MO imaging. In: MO imaging (eds Johansen T H and Shantsev D V) Kluwer: Dordrecht p 242
[22] Bechtle S, Ang S F and Schneider G A 2010 On the mechanical properties of hierarchically structured biological materials. *Biomaterials* **31** 6378

IOP Conf. Series: Materials Science and Engineering **625** (2019) 012007 doi:10.1088/1757-899X/625/1/012007

Development of a stretched absolute wavelength standard using dichroic atomic vapor spectroscopy towards observation of light-induced drift

Y Kusano[1], N Nishiya[1] and L Matsuoka[1]

[1] Graduate School of Engineering, Hiroshima University, Higashi-Hiroshima, Hiroshima 739-8527, Japan

Email: m181737@hiroshima-u.ac.jp

Abstract. An absolute wavelength standard broader than the Doppler width of the atomic absorption signal was developed by using a method called dichroic atomic vapor spectroscopy (DAVS). DAVS is known as a method to obtain an error signal for stabilization of laser wavelength, called dichroic atomic vapor laser locking (DAVLL). In the DAVS, a magnetic field is applied to the atomic vapor cell and a stretched signal is obtained by the Zeeman effect. The DAVS experiment was performed by using ring-type permanent magnets and a cesium vapor cell at room temperature, and the placement and number of magnets were optimized. By measuring the height of the DAVS signal calibrated in advance, continuous monitoring of the absolute laser wavelength become possible. The developed system is suitable to be used as a stretched wavelength standard in the experiment of the light-induced drift.

1. Introduction

The radioactive wastes generated in nuclear power generation are planned to be solidified in the glass and disposed at deep underground. However, there are various social problems such as the location determination for the final disposal site and the emission risk due to the long-term storage. The nuclear transmutation of the isotopes with a long half-life is one possibility to reduce the geological disposal. For the transmutation of Cs-135, which is a dominant isotope from the viewpoint of the public exposure, the almost complete isotope separation from Cs-133 is required [1]. However, the efficient separation of Cs isotopes has not been realized because the mass of the Cs isotopes are very close and other properties such as the electronic configurations are almost the same. The isotope separation by the light-induced drift (LID) [2-4] has a possibility to be extended to the industrially effective level because the separation principle works in the thermal equilibrium. The theoretical investigation of the separation of cesium isotopes by LID were performed previously, and resulted in hopeful speculations [5].

In the LID experiment, by optically exciting only the atoms of a certain velocity, the drift velocity is driven by the velocity-selective excitation and the difference of the redistribution speed among the electronic states. For the experiment of the LID, it is necessary to make a wide range absolute wavelength standard because we have to monitor the frequency of the laser to ascertain which atom of the velocity component near the resonant line of cesium is excited by the laser. The drift velocity emerges around ±1 GHz from the center frequency of the target transition and the direction of the velocity is steeply changing around the center frequency. The continuous and stable wavelength standard is required to obtain the dependence of drift velocity on the laser frequency.

Generally, the wavelength standard for laser spectroscopy is created by the atomic transition lines, the interferometer, or combination of them. However, the atomic transition line is stable but not continuous whereas the signal of the interferometer is continuous but hard to be stable. For a stable and continuous wavelength standard, we apply the dichroic atomic vapor spectroscopy (DAVS), which is used for the wavelength stabilization called dichroic atomic vapor laser locking (DAVLL). The works of DAVLL are roughly classified into Doppler DAVLL [6-12] or sub-Doppler DAVLL [13-20]. The sub-Doppler DAVLL has the advantage of locking the frequency inside the narrow region. In the present study, only the Doppler version of DAVS was utilized because the signal over the broad frequency range was important. The similar interest for the wide range signal creation was presented recently [21]. The DAVS provides us the time-stable signal at room temperature by a relatively simple experimental apparatus.

In this study, the signal of DAVS was created by using ring-shaped permanent magnets and cesium vapor cell. Since the width of the DAVS signal can be changed by the intensity of the magnetic field and how to use the magnets, the conditions of the magnet arrangement were optimized to be used as a wide range absolute wavelength standard.

2. Experiment

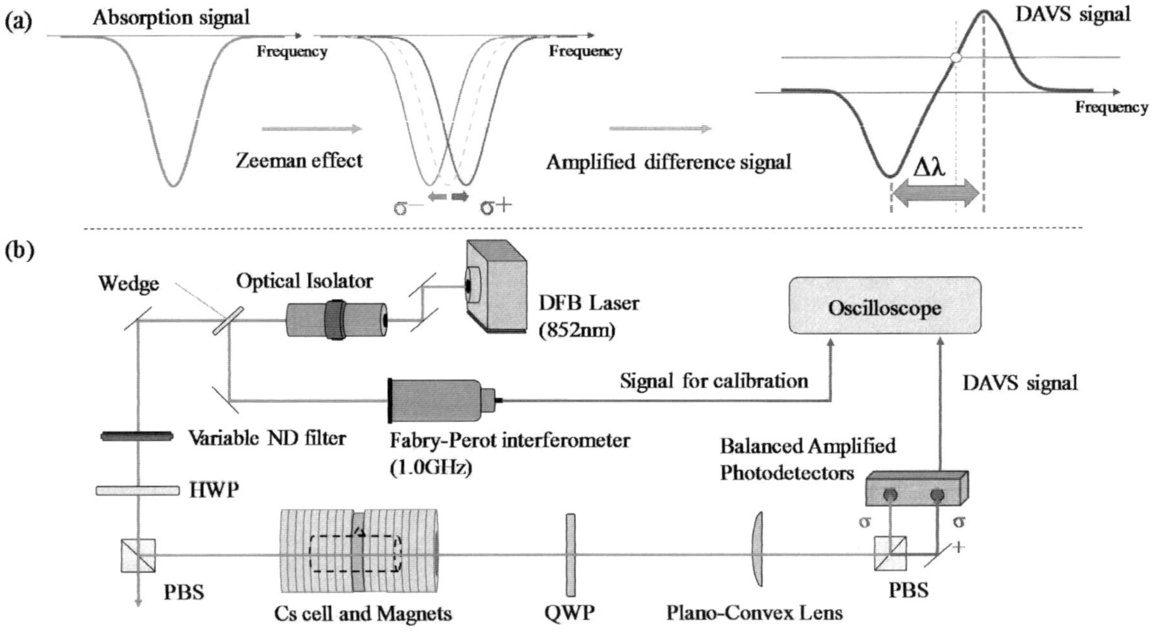

Figure 1. The schematic of the DAVS signal creation (a) and the experimental system (b). HWP, QWP, and PBS mean half-wave plate, quarter-wave plate, and polarizing beam splitter, respectively.

2.1. Principle

The DAVS is often used for making a reference error signal for wavelength stabilization of diode lasers. The laser locking using DAVS has an advantage for providing absolute wavelength reference, stability for temperature change and high-speed feedback because of no need for lock-in amplification. In addition, we made use of the feature of the shape of the DAVS signal broadened around a few GHz.

Figure 1 (a) shows the principle of the DAVS signal creation. Under the magnetic field, the transition wavelengths of cesium atoms are separated and shifted by the Zeeman-effect. The direction of the shifts depends on the circularly-polarized component of the laser. Because the linearly polarized light is expressed by the linear combination of the circularly-polarized lights which have opposite chiralities, both shifts are simultaneously observed by the linearly-polarized laser. By using the quarter-wave plate,

IOP Conf. Series: Materials Science and Engineering **625** (2019) 012007 doi:10.1088/1757-899X/625/1/012007

the transmitted light intensity ratio of the circularly-polarizing components is transferred to the ratio of horizontally and vertically polarizing components, which can be easily separated by the polarizing beam splitter. The DAVS signal is composed by the difference of two signals given by each polarizing component.

2.2. Experimental apparatus

Figure 1 (b) is the schematic of the experiment. A distributed feedback laser diode (LD-0852-0150-DFB-1, Toptica) was mounted on the TO-3 laser mount (246, Arroyo Instruments) connected to the current and temperature controller (6305, Arroyo Instruments). The wavelength of the laser was scanned by modulating the injecting current using a function generator (WF1974, NF Corporation). The optical isolator (IOT-5-850-MP, Thorlabs) was used to prevent the back reflection causing the instability of wavelength. The power of the laser was attenuated by a pair of the half-wave plate and the polarizing beam splitter. The linearly-polarized laser was inserted into the cesium gas cell composed of Pyrex (length of 100 mm and diameter of 30 mm, Horizon). To generate the magnetic field, the ring-shaped permanent magnets (thickness of 8 mm) were used. The average magnetic field strength of one magnet measured in the center of the ring was 356.5 Gauss. The output from the cell was separated into the two components by a polarizing beam splitter after the quarter-wave plate that transfers the ratio of circular-polarization into the ratio of horizontal and vertical linear polarization. The DAVS signal was generated by measuring the intensity of two components with a balanced amplified photodetector (PDB410A, Thorlabs). A plano-convex lens with a focal length of 150 mm is used to focus the laser into the detector.

The DAVS signal is obtained on the oscilloscope (TDS 8204, Owon) with the signal of a Fabry-Perot Interferometer (FPI 100-0750, Toptica photonics, FSR: 1.0 GHz) for relative frequency calibration.

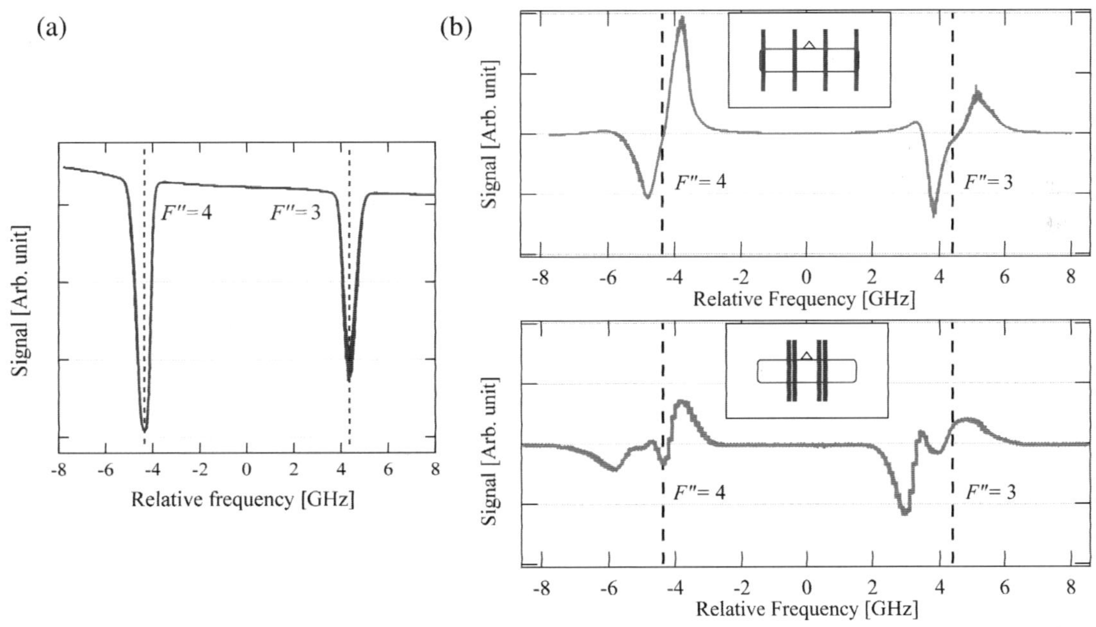

Figure 2. An example of the absorption signal of the cesium atoms without ring magnets (a), examples of the DAVS signal depending on the position of four ring-shaped magnets (b). Inset illustrations are examples of how to place 4 magnets. The upper shows a DAVS signal when the magnets arranged at equal intervals, and the lower shows those are arranged non-uniformly.

3. Results and discussion

Figure 2(a) shows an absorption signal of cesium atoms in the region of D_2 line. In this region, two peaks given by the hyperfine splitting of the ground state are observed. Each peak involves some transitions given by the hyperfine splitting of the excited state.

Figure 2(b) shows an example of the DAVS signal depending on the position of four magnets. If the magnets are put around the center without the spacer between each pair, the DAVS signal did not exhibit the single slope structure ideal for the wavelength reference. This is because the direction of the magnetic field varies inside the cell. The magnetic fields inside the center hole of the ring magnet and outside it are opposite, resulting in the unwanted splitting of the cesium signal due to the variation of the magnetic field. If the magnets were placed at both ends of the cell and nearly even spaces for others, the single slope structure was obtained because the direction of the magnetic field is nearly the same inside the cell in this case.

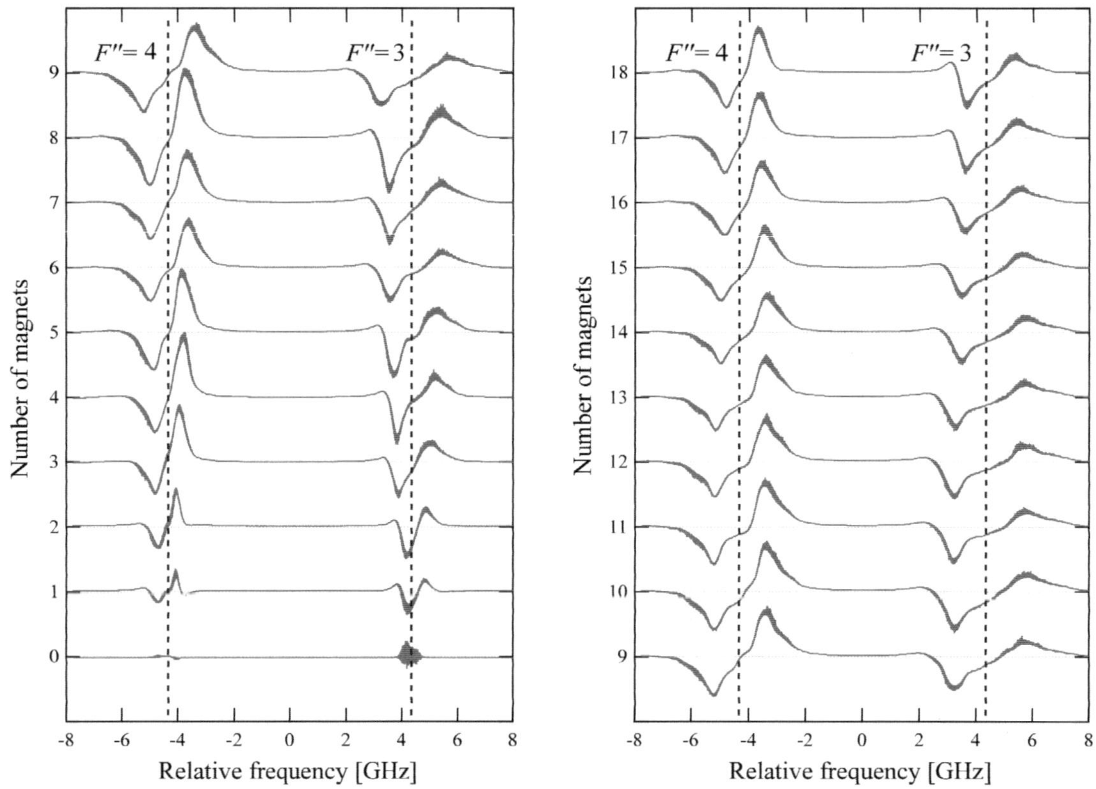

Figure 3. The DAVS signals obtained by varying the number of ring magnets.

Figure 3 shows the variation of the DAVS spectra depending on the number of ring magnets. The magnets were placed with nearly even spacing. The width of the slope signal got wider with increasing the number of magnets until 10 and started to decrease gradually. With 10 magnets, the cesium cell was just completely covered by the magnets. For more than 11 magnets, the magnets are stuck out. The extra magnets worked to decrease the magnetic field inside the cell.

As the wavelength reference, a monotonically increasing signal was obtained over 2 GHz, which is sufficient to be used in the experiment of the LID. Because the DAVS signal is based on the atomic absorption signal, the signal level and the absolute frequency of the laser can be connected with one-to-one correspondence without fluctuation. And that is relatively robust to the temperature change.

4. Conclusion

The monotonically changing in frequency and stable in time absolute wavelength standard was obtained as the signal of the DAVS. To obtain the signal, it was appropriate to place permanent magnets at equal intervals as much as possible around the Cs cell. The width of the slope of the signal was over 2 GHz and was enough larger than the Doppler width of cesium. It can be used for monitoring the wavelength of lasers in the LID experiment in the future.

Acknowledgments

This work was supported by JSPS KAKENHI Grant Number JP16K14538, JP17H04978, and TEPCO Memorial Foundation Research Grant (Basic Research) H28.

References

[1] Chaltikyan V, Papoyan A, Oshita H, Shiotani H, Ono K, Ishikawa M and Ozawa M 2009 *J. Radioanal. Nucl. Chem.* **280** p 343
[2] Gelmukhanov F K and Shalagin A M 1979 *JEPT Lett.* **29** p 711
[3] Eliel E R 1992 *Adv. At. Mol. Opt. Phys.* **30** p 199
[4] Sato S and Saito M 1996 *Mater. Trans. JIM.* **37** p 1789
[5] Yuki K, Kobayashi T and Matsuoka L 2017 *JOURNAL OF NUCLEAR SCIENCE AND TECHNOLOGY* **54** p 1240
[6] Corwin K L, Lu Z T, Hand C F, Epstein R J and Wieman C E 1998 *APPLIED OPTICS* **37** p 3295
[7] McCarron D J, Hughes I G, Tierney P and Cornish S L 2007 *Rev. Sci. Instrum.* **78** 093106
[8] Hasegawa T and Deguchi M 2009 *J. Opt. Soc. Am. B* **26** p 1216
[9] Lee C, Iwata G Z, Corsini E, Higbie J M, Knappe S, Ledbetter M P and Budker D 2011 Rev. Sci. Instrum. 82 043107
[10] Yin S, Liu H, Qian J, Hong T, Xu Z and Wang Y 2012 *Opt. Commun.* **285** p 5169
[11] Xiao T, Wang T and Yan B 2018 *Rev. Sci. Instrum.* **89** 046106
[12] Su Dian-Qiang, Liu Rui-Juan, Zhang Chuan-Biao, Ji Zhong-Hua, Zhao Yan-Ting, Xiao Lian-Tuan and Jia Suo-Tang 2018 *J. Phys. D: Appl. Phys.* **51** 465001
[13] Wasik G, Gawlik W, Zachorowski J and Zawadzki W 2002 *Appl. Phys. B* **75** p 613
[14] Petelski T, Fattori M, Lamporesi G, Stuhler J and Tino G M 2003 *Eur. Phys. J. D* **22** p 279
[15] Okubo S, Iwakuni K and Hasegawa T 2012 *Opt. Commun.* **285** p 4107
[16] Harris M L, Cornish S L, Tripathi A and Hughes I G 2014 *J. Phys. B* **41** 085401
[17] Su Dian-Qiang, Meng Teng-Fei, Ji Zhong-Hua, Yuan Jin-Peng, Zhao Yan-Ting, Xiao Lian-Tuan and Jia Suo-Tang 2014 *APPLIED OPTICS* **53** p 7011
[18] Choi G W and Noh H R 2015 *J. Phys. B* **48** 115008
[19] Choi G W and Noh H R 2016 *Opt. Commun.* **367** p 312
[20] Singh V, Tiwari V B, Mishra S R and Rawat H S 2016 *Appl. Phys. B* **122** 225
[21] Pustelny S, Schultze V, Scholtes T and Budker D 2016 *Rev. Sci. Instrum.* **87** 063107

IOP Conf. Series: Materials Science and Engineering **625** (2019) 012007 doi:10.1088/1757-899X/625/1/012007

First name	Surname	Institution	Email address	Title (Prof/ Dr./student)	Research Interests
Yuya	Kusano	Graduate School of Engineering, Hiroshima University, Higashi-Hiroshima, Hiroshima 739-8527, Japan	m181737 @hiroshima-u.ac.jp	Student	Mechanical engineering and Optical engineering
Noritaka	Nishiya	Graduate School of Engineering, Hiroshima University, Higashi-Hiroshima, Hiroshima 739-8527, Japan	m171318 @hiroshima-u.ac.jp	Student	Atomic spectroscopy and Swarm intelligence
Leo	Matsuoka	Graduate School of Engineering, Hiroshima University, Higashi-Hiroshima, Hiroshima 739-8527, Japan	leo-matsuoka @hiroshima-u.ac.jp	Assistant Prof.	Isotope separation and Atomic physics

IOP Conf. Series: Materials Science and Engineering **625** (2019) 012008 doi:10.1088/1757-899X/625/1/012008

Splitting tensile strength of lightweight concrete using polypropylene coarse aggregate coated with sand

H Purnomo[1], R K Putri[1] and P Perdani[1]

[1] Civil Engineering Department, Universitas Indonesia, Depok 16424, Indonesia

Email: herupur@eng.ui.ac.id

Abstract. As part of efforts to reduce plastic waste, transformation of it to one concrete component will be good for the environment. The use of plastic waste as coarse aggregate in concrete making can provide lighter concrete but with a lower strength compared to normal concrete. Accordingly, an experimental study of 42 concrete specimens using waste polypropylene coarse aggregates coated with sand was carried out. Splitting tensile tests were conducted to cylinder concrete specimens having diameter of 15 cm and depth of 30 cm respectively. Three mixtures of sand coated polypropylene coarse plastic aggregate, river sand as fine aggregate, water and Portland Composite Cement with a water-cement ratio of 0.29, 0.287 and 0.255 were conducted. The mass proportion of cement and sand are the same but the mass of plastic coarse aggregates coated with sand is specific for each mixture. Splitting tensile strength of the specimens in general shows that higher tensile strength is found for specimens having higher compressive strength. From the test results, two equations of splitting tensile strength are proposed where it is based on three different quality of lightweight concrete at the age of 7 and 28 days. Finally the splitting tensile strength for lightweight concrete using polypropylene coarse aggregate coated with sand is found to be higher than the direct tensile strength for the same lightweight concrete.

1. Introduction

Recent studies confirm the use of plastic aggregates as component of concrete material with satisfactory results [1, 2]. The use of it as one of the component of a construction material could minimize plastic waste problem. As there are many kind of plastic waste from industries, there is lack of mechanical properties information of lightweight concrete using a specific plastic waste aggregate. Beside compressive strength, shear strength, elasticity modulus and Poisson ratio, tensile strength is considered as one of the important mechanical properties in the civil engineering domain. Direct tension, splitting and flexure tests are often used for determining the tensile strengths of normal and lightweight concrete. Flexure and splitting concrete tensile strength are considered higher in values compared to direct tensile strength [3, 4]. Since there are three formulations for concrete tensile strength, appropriate engineering judgement should be used in the design and construction of civil engineering concrete structures. Recent studies of splitting tensile test of concrete could give new perspective in the construction industry. In this regard, Yan K et al [5] have presented a possible applicability of support vector machine (SVM) to predict the splitting tensile strength of concrete from compressive strength of concrete. Another study [6] shows a better understanding of the effects of a construction joint on the splitting tensile strength could reducing the amount of steel reinforcement at the construction joints.

In this study, splitting tensile strength of lightweight concrete using polypropylene waste plastic coarse aggregate which is coated with sand is obtained as function of the compressive strength at age of 7 and 28 days. Finally the splitting tensile strength equation is compared to direct tensile strength from previous study.

2. Experimental program

The main objective of this study is to obtain the splitting tensile strength of the lightweight concrete. A series of trial mix was conducted previously before conducting the compression and splitting tensile tests. The preparation of the material needed for concrete mix design is described in the following paragraph.

2.1 *Materials*

Portland Cement Composite was used to comply Indonesian Standard (SNI 15-7064-2004) [7] and British Standard (BS EN 197-1:2000) [8]. The coarse aggregate was produced from waste of Polypropylene (PP) plastic. Figure 1 presents the original shapes of uncoated plastic aggregate. The uncoated aggregate shape is similar to the plastic aggregate designed by Pamudji, G et al [9]. The size of the biggest aggregate has generally 10 mm of thickness and, 20 mm of length and 20 mm of width. The coated plastic aggregate was produced by the same method as presented in [10]. In this regard, plastic aggregate was established by heating the plastic up to the melting point, then flowed into the mould by hydraulic pressure. The process itself was conducted using manual injection plastic machine with automatic temperature control. The melting temperature of the plastic is in range of $130^0C \pm 10^0C$. The plastic aggregate were then cooled after the formation. Afterward it was coated with hot sand by placing the uncoated plastic aggregate into a rotating cylinder. The final shape of plastic aggregate coated with sand is presented in Figure 2. The coating is expected to improve bonding at the interface between coarse aggregate and mortar. Abrasion test was established afterward, and the coarse aggregate abrasion is obtained not more than 10.8%.

Figure 1. Shape of uncoated plastic aggregate [9]

Figure 2. Plastic aggregate coated with sand [10]

Sand was taken from Cimangkok which is located at West Java Province, Indonesia. It was used as fine aggregate. Size of the fine aggregate is less than 9.5 mm while its bulk specific gravity and fineness modulus of the fine aggregate are found to be 2.588 and 2.316 respectively. The passing percentage of fine aggregate is within the limits as indicated in Indonesian standard SNI 03-2834-2000 [11].

2.2 *Mix proportion of concrete*

Proportion of materials was determined using mix design formulation that is based on absolute volume. From trial mixes, 3 (three) material mix proportion are denoted by mix design or mixture (MD1. MD2 and MD3) have fulfilled the material mixture for structural purposes. For the 3 mixtures, the results of the sieve analysis of the coarse aggregate are almost the same as shown in Figure 3a and 3b, where the results are slightly outside the lower and upper limit as indicated in Indonesian Standard SNI 03-2834-2000. The sieve analysis of the fine aggregate is presented in Figure 4 where the result is within both limit, upper and lower.

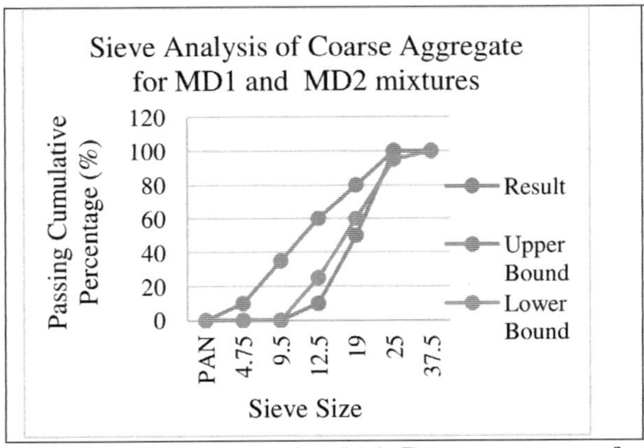

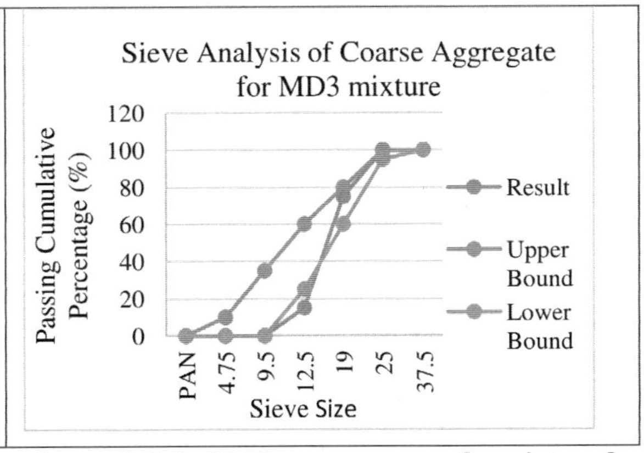

Figure 3. Sieve analysis of (a) Coarse aggregate for mixture 1 & 2, (b) Coarse aggregate for mixture 3

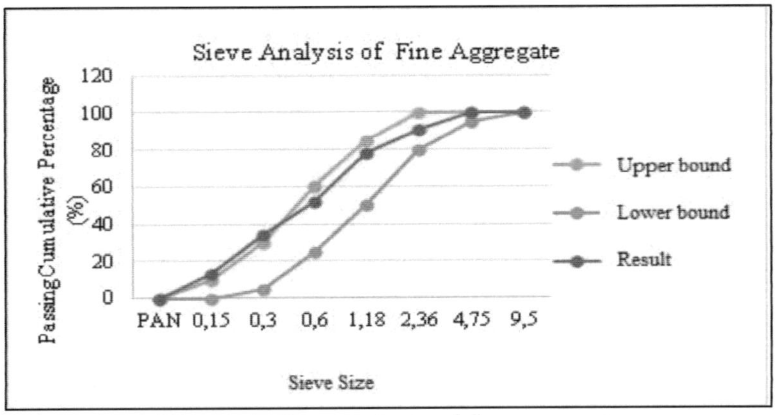

Figure 4. Sieve analysis of fine aggregate

Similar water to cement ratio (W/C) were used for the 2 (three) mixtures, which are 0.290 and 0.287 for mixtures MD1 and MD2 respectively, while mix design MD3 has 0.256 as presented in Table 1. Superplasticizer (SP) was used in order to have better workability to mix all materials of MD1 to MD3.

Table 1. Concrete mix proportion

Mixture Type	W/C	Cement (kg)	Sand (kg)	Coarse Plastic Aggregate (kg)	Water (kg)	SP (kg)
MD1	0.290	482	826.0	434.686	139	2.90
MD2	0.287	529	909.9	367.773	152	3.70
MD3	0.256	558	961.0	349.707	143	4.46

2.3 Density of concrete specimens

The splitting tensile test of concrete specimens was carried out on 150 mm diameter and 300 mm height cylinder specimens where at 7 days 3 (three) specimens , and at 28days 4 (four) cylinders were tested for each mixture. The density of these 21 cylinders specimens are listed in the Table 2. The densities of other 21 cylinders specimens for compression test are found almost in the similar range as shown in Table 2.

IOP Conf. Series: Materials Science and Engineering **625** (2019) 012008 doi:10.1088/1757-899X/625/1/012008

Table 2. Average concrete density

Age (days)	MD1 (kg/m3)	MD2 (kg/m3)	MD3 (kg/m3)	Number of specimen
7	1780.48	1785.47	1811.75	9
28	1738.22	1779.77	1781.27	12

2.4 Compression test of concrete specimens

The compression test of concrete specimens was carried out on 150 mm diameter and 300 mm height cylinder specimens where at 7 days 3 (three) specimens were tested, and at 28 days 4 (four) cylinders were compressed for each mixture . Compression test was done by following ASTM C-39 [12]. The compression test was conducted at the 7th and 28th day for a total of 21 specimens tested as shown in the Table 3.

Table 3. Average concrete compressive strength

Age (days)	MD1 (MPa)	MD2 (MPa)	MD3 (MPa)	Number of specimen
7	17.03	19.28	20.34	9
28	21.01	23.45	26.58	12

2.5 Splitting tensile test of concrete specimens

The splitting tensile test of concrete specimens was carried out on 150 mm diameter and 300 mm height cylinder specimens where at 7 days 3 (three) specimens were tested, and at 28days 4 (four) cylinders were examined for each mixture . Splitting tensile test was carried out by following ASTM C-496 [13]. The test was conducted at the 7th and 28th day for a total of 21 specimens tested as shown in the Table 4.

Table 4. Average concrete splitting tensile strength

Age (days)	MD1 (MPa)	MD2 (MPa)	MD3 (MPa)	Number of specimen
7	1.42	2.06	2.39	9
28	2.12	2.43	2.52	12

2.6 Splitting tensile and compressive strength relation

Having all the average compressive and splitting tensile strength at age of 7 days and 28 days for all the 3 mixtures as shown in the Table 5 and Table 6, we can find the relation between the concrete splitting tensile f_{sp} and compressive strength $f_{c'}$ for 7 days and 28 days old concrete as presented by the equation (1) and (2).

Table 5. 7 days splitting tensile and compressive strength

Mixture type	Compressive strength (MPa)	Splitting tensile strength (MPa)	Percentage to compressive strength (%)	Average percentage (%)
MD1	17.03	1.42	8.34	10.256
MD2	19.28	2.06	10.68	
MD3	20.34	2.39	11.75	

$$f_{sp} = 0{,}1025\ fc' \quad \text{for concrete age of 7 days} \tag{1}$$

Table 6. 28 days splitting tensile and compressive strength

Mixture type	Compressive strength (MPa)	Splitting tensile strength (MPa)	Percentage to compressive strength (%)	Average percentage (%)
MD1	21.01	2.12	10.09	9.977
MD2	23.45	2.43	10.36	
MD3	26.58	2.52	9.48	

$$f_{sp} = 0{,}0997\ fc' \quad \text{for concrete age of 28 days} \tag{2}$$

3. Splitting tensile and direct tensile strengths

Previous study [14] has found the direct tensile strength $\boldsymbol{f_t}$ as function of the compressive strength from the same type of lightweight concrete, which is presented by the equation (3).

$$f_t = 0{,}42\ \sqrt{fc'} \quad \text{for concrete age of 28 days} \tag{3}$$

By introducing the same compressive strength into the equation (2) and (3), we can obtained a lower value of direct tensile strength compared to that of splitting tensile strength for the same specific lightweight concrete material. Similar situation was also found by Kim and Taha in their study [15].

4. Failure pattern of concrete specimen

From 21 specimens tested, all specimens have experienced failure almost at the middle part of the concrete cylinder specimens. Example of this typical failure is shown in the Figure 5 for concrete specimen at age of 7 days, and Figure 6 represents typical failure from specimen at the age of 28 days.

5. Conclusive remarks

In this study, it is concluded that splitting tensile strength is dependent on the compressive strength of the concrete. The higher the compression strength the higher the splitting tensile strength will be. The splitting tensile strength of lightweight concrete studied is higher compared to that of direct tensile strength obtained from previous study. This study also reveals that polypropylene coarse aggregate coated with sand contributes in fulfilling the lightweight structural concrete criteria. Two equations of splitting tensile strength are proposed where it is based on three different quality of lightweight concrete at the age of 7 and 28 days. The failure pattern of all specimens, are typical where it is located at the middle part of the concrete cylindrical specimen following the direction of the vertical load applied to the specimens. Future work involving flexure tests of beams made of lightweight

IOP Conf. Series: Materials Science and Engineering **625** (2019) 012008 doi:10.1088/1757-899X/625/1/012008

concrete using polypropylene coarse aggregate coated with sand are to be conducted to have insight of the tensile strengths from direct tension, splitting and flexure tests.

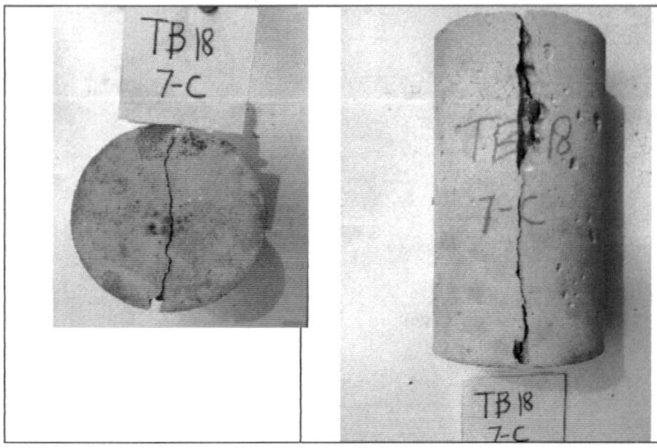

Figure 5. Typical failure of 7 days old specimen

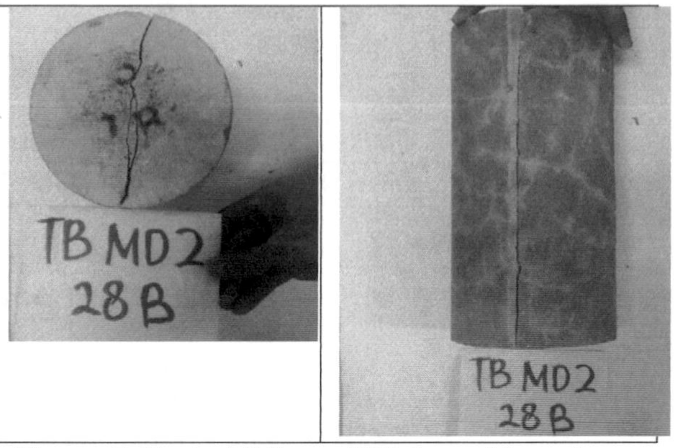

Figure 6. Typical failure of 28 days old specimen

Acknowledgement
This study was supported by the Ministry of Research Technology and Higher Education through the University Prime Basic Research Grant Funding Support. The authors also conveyed their gratitude to the staffs of the Laboratory of Structure and Materials, Department of Civil Engineering, Faculty of Engineering, Universitas Indonesia for their support in the experimentation program.

References
[1] Choi Y W, Moon D J, Kim Y J and Lachemi M 2009 *Constr. Build. Mater.* 23 pp 2829–35
[2] Saikia N and de Brito J 2012 *Constr. Build. Mater.* 34 pp 385–401
[3] Wu M, Zang C and Chen Z 2015 *J. Hydraul. Eng.* 46(8) pp 981-988
[4] Khan M I 2011 *Appl. Mech. Mater.* 117-119 pp 9-14
[5] Yan K, Xu H, Shen G, and Liu P 2013 *Adv. Mater. Sci. Eng.* ID 597257 pp 1-13
[6] Gerges N N, Issa C A, Fawaz S 2015 *Case Studies in Constr. Mater.* 3 pp 83-89
[7] SNI 15-7064-2004 *Standar Nasional Indonesia.* Semen Portland Komposit. Badan Standardisasi Nasional.
[8] BS EN 197-1:2000 Cement. Composition, specifications and conformity criteria for common cements ISBN 9780580580451
[9] Pamudji G., Purnomo H., Katili I and Imran I 2013 Proc. the 6th Civil Engineering Conf. in Asia Region (Jakarta) ISBN 978-602-8605
[10] Purnomo H, Pamudji G and Satim M 2017 MATEC Web of Conference 101, 01016
[11] SNI 03-2834-2000. *Standar Nasional Indonesia.* Tata Cara Pembuatan Rencana Campuran Beton Normal. Badan Standardisasi Nasional.
[12] ASTM C 39/C 39M-04a 2004 Standard Test Method for Compressive Strength of Cylindrical Concrete Specimens, American Standard Test Method, West Conshohocken, Penn, USA.
[13] ASTM C 496/C 496M-04 2004 Standard Test Method for Splitting Tensile Strength of Cylindrical Concrete Specimens, American Standard Test Method, West Conshohocken, Penn, USA.
[14] Purnomo H, Yuandry R A and Tjahjono E 2018 *Key Eng. Mater.* 789 pp 131-136
[15] Kim J J and Taha M 2014 *Adv. Civil* Eng. Article ID 156926 pp 1-8

IOP Conf. Series: Materials Science and Engineering **625** (2019) 012009 doi:10.1088/1757-899X/625/1/012009

The effect of the composition of magnesia flux on the sinter structure and properties

A N Shapovalov[1], E V Ovchinnikova[1], V B Gorbunov[1], R R Dema[2] and O B Kalugina[2,3]

[1] National University of Science and Technology «MISIS», Moscow, Russia
[2] Nosov Magnitogorsk State University, Magnitogorsk, Russia

Email: alshapo@misis.ru

Abstract. The metallurgical properties of magnesian fluxes of the Khalilovo deposit with different ratios in the composition of magnesite and serpentine are studied. The results of laboratory study experiments the effect of magnesian fluxes properties of the Khalilovo deposit on the sinter production from the ore of the Kursk Magnetic Anomaly at the plant of JSC Ural Steel are presented. It is revealed that the use of experienced fluxes contributes to the sinter strength, increase the yield ratio and specific capacity. Thus, using experienced fluxes instead of Bakal siderite, an increase in the yield of sinter is provided by 3-5 % (rel.) as a result of acceleration of solid-phase reactions involving magnesite and serpentine. At the same time, the sinter productivity increases from 1.04 to 1.08-1.15 t /m^2/h, that is, by 4-10% (rel.). The use of experienced magnesian fluxes supports to increase the drum strength of the sinter by 4-6 % abs. Improving the strength characteristics of the sinter when using the magnesian fluxes of the Khalilovo deposit is due to the formation of a "reinforcing" of hedenbergite solid solutions as well as homogenization of the solidifying melt and crystallization of it in the form of a glass phase of rankinite composition, which together limit the formation of β-Ca_2SiO_4.

1. Introduction

One of the main ways to increasing the efficiency of blast-furnace smelting is the optimization of the basicity and the magnesia content in the slag [1-6], the correction of which is carried out by changing the composition of the sinter, as a rule. In the practice of sintering production, various materials based on magnesite, dolomite, olivine, dunite, serpentine, etc. are used as sources of magnesia [7-12]. The choice of magnesia flux is made on the basis of its chemical and mineralogical composition, properties and transport accessibility.

As a magnesian component in the production of iron sinter at the plant of JSC Ural Steel, siderite ores of the Bakal deposit with a content of 30-32 % Fe and 10-12 % MgO are used. The main ore-forming mineral of Bakal siderite is sideroplesite - (Mg, Mn, Fe) CO_3), its percent is 75-80%. The remaining 20-25 % are dolomite, silicates in the form of shale, quartz, opal, etc. This material has an unstable chemical composition and a high content of large fractions (more than 10 mm), which adversely affects the agglomeration process and the quality of the sinter. At the same time, near from JSC Ural Steel, the Khalilovo serpentine-magnesite deposit with a content of MgO up to 40 % is being developed. The main ore-forming mineral of the Khalilovo deposit is serpentine ($Mg_6[Si_4O_{10}(OH)_2]OH_6$), its ratio in the ore is 55-90 %; carbonates — dolomite (10–14 %) and magnesite (5–40 %) are concomitant. In order to estimate the effectiveness of using magnesian fluxes from serpentine-magnesites of the Khalilovo deposit for sinter production, we conducted a study on the effect of magnesian flux (with different magnesite content) on the agglomeration process and sinter quality.

2. Materials and methods

Fluxes from serpentine-magnesites of the Khalilovo deposit with different magnesite contents of 10, 30, 50, and 70 % were used as experimental materials. This choice is due to the current multi-stage technology for the enrichment of serpentine-magnesites, which makes it possible to obtain a product with different magnesite content at various stages of processing. The chemical compositions of the experimental fluxes, as well as the Bakal siderite, are given in table 1.

Table 1. Chemical composition of experimental magnesian materials

Component title	The component's content of the material,%				
	Bakal siderite	Magnesia flux of the Khalilovo deposit with different magnesite percentage (%)			
		10	30	50	70
Fe	30,2	7,59	6,71	5,82	4,93
FeO	32,98	0,71	0,55	0,39	0,24
SiO_2	7,58	31,05	24,15	17,25	10,35
CaO	4,22	2,22	1,75	1,28	0,81
Al_2O_3	1,9	0,88	0,69	0,49	0,29
MgO	10,4	38,22	39,75	41,28	42,82
S	0,17	0,027	0,021	0,015	0,009
P_2O_5	0,057	0,144	0,112	0,080	0,048
Ignition losses	34,53	16,58	24,93	31,27	38,61

The studies were conducted in the laboratory of JSC "Ural Steel" on a sinter pot with a diameter of 210 mm. Mixing, moistening (up to 7.0–7.5 %) and pelletizing of the components of the sinter mix were carried out in a drum granulator 600 mm in diameter for 5 minutes with a rotation speed of 12 rpm. Iron ore concentrate and sintering ore of Mikhailovsky GOK were used as the main components of the sinter mix. The sinter mix was calculated in order to obtain the sinter with a CaO / SiO_2 basicity of 1.5 units with a MgO content of 2.0 %. The required a MgO content in the sinter was ensured by the expenditure of Bakal siderite (in the base period) and experimental fluxes with different magnesite content (in the experimental periods). Maintaining the basicity at a required level was provided by adjusting the consumption of limestone. The content of return fines in the sinter mix was 25 %, and carbon it was 4.2 %.

3. Experiment

The averaged results of the experimental sintering are presented in table 2.

Table 2. Average results of experiments

Component title	The results of experiments using different sources of MgO				
	Bakal siderite	Magnesia flux with different magnesite percentage (%)			
		10	30	50	70
Sinter mix moisture, %	7,10	7,20	7,10	7,20	7,30
Height of the sintered layer, mm	347	346	345	346	344
Sintering rate, mm / min	15,95	16,16	16,28	16,28	16,36

IOP Conf. Series: Materials Science and Engineering **625** (2019) 012009 doi:10.1088/1757-899X/625/1/012009

The yield of sinter +5 mm from sinter cake,%		76,90	77,93	77,96	78,30	78,01
Productivity, t/m^2/hour		1,04	1,08	1,13	1,15	1,14
Abrasion resistance (Russian all-Union State Standard 15137-77),%		5,38	5,40	4,45	4,55	4,58
Tumbler index (Russian all-Union State Standard 15137-77), %		65,68	68,30	71,08	72,65	71,98
Sinter composition, %	Fe	53,20	52,85	53,25	53,45	53,60
	FeO	12,20	12,85	12,60	12,90	13,15
	SiO_2	9,00	9,20	9,05	9,05	8,90
	CaO	13,10	13,20	13,35	13,65	12,95
	MgO	1,90	2,00	1,92	2,01	2,09
	S	0,017	0,019	0,015	0,019	0,020
Basicity CaO/SiO_2		1,46	1,44	1,48	1,51	1,46

Comparison of the experimental results allows us to conclude that the sintering process using the experimental magnesian fluxes exceed those of the Bakal siderite. The use of experienced magnesia fluxes provides an increase in productivity by sintering rate increase and by a step-up of sinter yield. An increase in the sinter yield (with a particle size of more than 5 mm) is probably associated with improved thermal sintering conditions (confirmed by an increase of FeO content in the sinter with constant fuel consumption), as well as a change in the mechanism of mineral formation.

When using experienced magnesian fluxes, the tumbler index (TI + 5mm) and abrasion index (AI-0.5 mm) increase were observed. The index values were determined in a rotating drum according to the Russian all-Union State Standard 15137-77. The reasons for the significant improvement in the strength of the sinter obtained using serpentine-magnesite were defined by a comparative analysis of the "base" and experimental sinter microstructures.

Microstructure of sinter

The X-ray spectroscopy analysis of experimental sinter revealed that MgO almost completely goes into the ore part (magnetite) in all cases. At the same time, the crystallization form of the silicate bond of ore grains varies depending on the type of magnesian material. For example, figure 1 shows the fragments of microstructures of the sinter obtained with Bakal siderite (Figure 1, a) and with serpentine-magnesite with a magnesite content of 10, 30 and 50 % (Figure 1, b-d).

Table 3 shows the results of x-ray analysis of the sinter phases obtained with different magnesian materials.

IOP Conf. Series: Materials Science and Engineering **625** (2019) 012009 doi:10.1088/1757-899X/625/1/012009

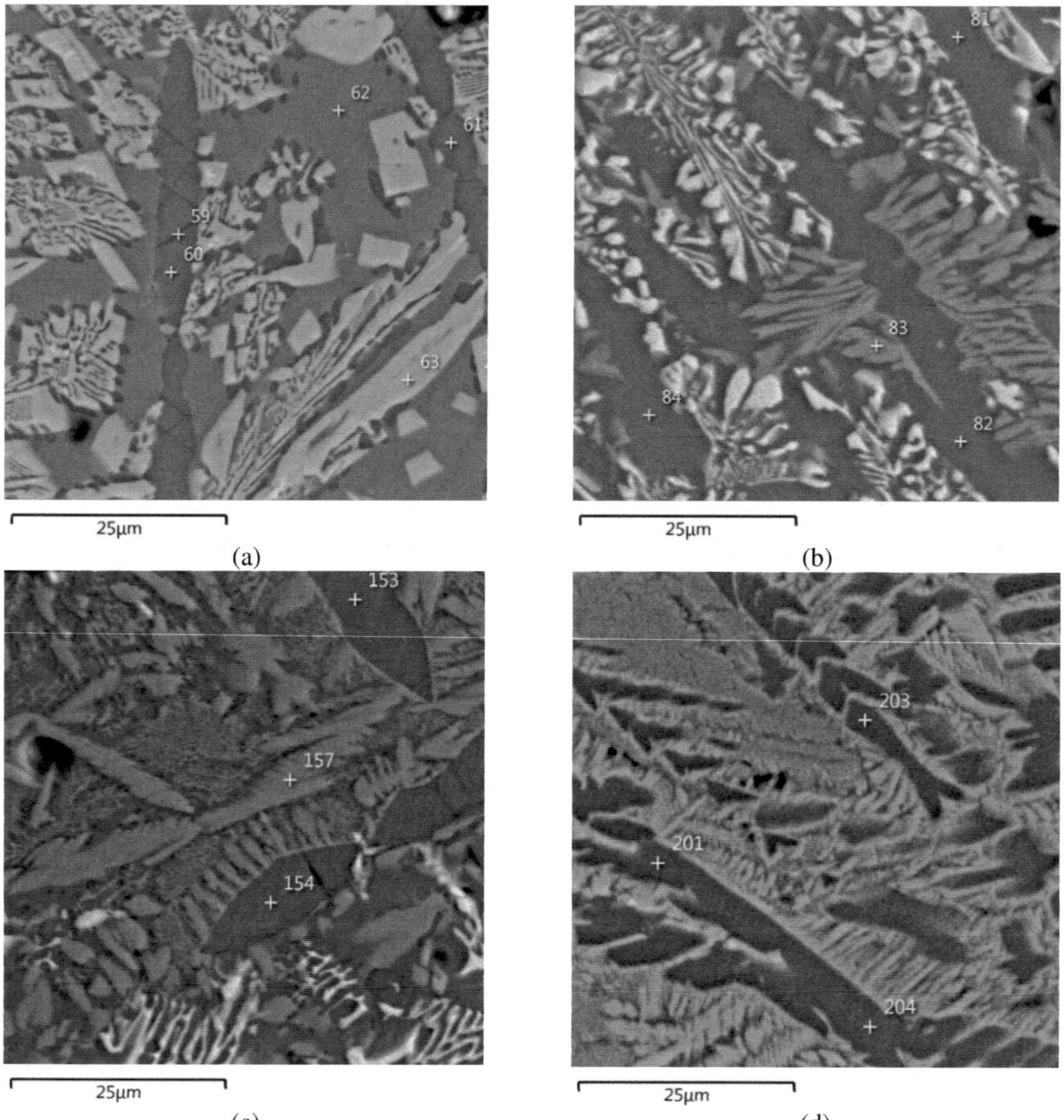

Figure 1. Sinter microstructure with using Bakal siderite (a) and serpentine-magnesite with 10 % magnesite (b), 30 % (c) and 50 % (d)

Table 3. Chemical composition of sinter ligament-phases with different magnesian fluxes (wt%)

Spectrum number (fig. 1)	O	Mg	Al	Si	Ca	Fe	Phase part
Sinter with Bakal siderite (Figure 1, a)							
59	37,97			16,2	44,63	1,03	Dicalcium silicate
60	38,02			16,44	43,76	1,32	Dicalcium silicate
61	37,64			16,36	44,3	1,47	Dicalcium silicate
62	38,74		2,45	18,55	24,74	12,33	Glass phase of low basic capacity

63	31,89	0,4	2,02	4,67	12,09	48,93	Calcium ferrite
Sinter with magnesia flux containing 10% magnesite (Figure 1, b)							
81	40,16			19,48	39,13	1,23	Glass phase of
82	39,87			19,84	39,24	1,05	rankinite composition
83	37,88		1,96	15,73	17,72	26,28	Mellit
84	40,5			19,56	39,02	0,92	Glass phase of rankinite composition
Sinter with magnesia flux containing 30% magnesite (Figure 1, c)							
153	37,84		2,15	16,52	44,69	0,71	Glass phase of
154	35,46			16,91	46,37	0,83	rankinite composition
157	35,99			14,38	18,64	28,39	Hedenbergite
Sinter with magnesia flux containing 50% magnesite (Figure 1, d)							
201	36,00	0,15		19,94	41,78	1,99	Glass phase of rankinite composition
203	35,93	0,19		19,59	40,96	3,33	
204	36,73	0,16		20,06	41,65	1,40	

An experimental sinter microstructure analysis allows us to conclude that the improvement in the strength of the sinter with serpentine-magnesite arises due to the simultaneous occurrence of the processes:

– formation of hedenbergite solid solutions (in the ferrites absence) [8, 13], which provides strengthening of the sinter cake;

– homogenization of the solidifying melt and its crystallization in the form of a glass phase of rankinite composition, which limits the formation of β-Ca_2SiO_4.

When a siderite from the Bakal deposit used as a magnesian material, the process of mineral formation is restrained by the lower activity of free MgO emerged during the decomposition of magnesium carbonate, as well as by the formation of refractory spinels of the magnesiomagnetite and magnesioferrite type [14-16].

Magnesium oxide, which is part of serpentin-magnesium-magnesite, is involved in the formation of silicate phases. The formation of refractory spinels at the MgO and iron oxides interaction takes placeonly at the stage of melt crystallization. The magnesite increase in the content of the serpentine-magnesite leads to an increase in the proportion of solid solutions of hedenbergite in the sinter.

4. Conclusions

The use of magnesia fluxes from serpentine-magnesites of the Khalilovo deposit contributes to the strengthening of the sinter, increasing the yield of the useful and specific productivity.

Improvement of the sinter strength characteristics and indicators of the sintering process occurs due to the formation of a "reinforcing" of hedenbergite solid solutions, as well as the homogenization of the solidifying melt and its crystallization in the form of a glass phase of rankinite composition, both of these processes together limit the formation of β-Ca_2SiO_4.

Acknowledgements

The work was financially supported by the Ministry of Education and Science of the Russian Federation, the project № 11.2054.2017/4.6 of the state assignment for 2017-2019.

References

[1] Grigor'ev F F, Chernyatin A N, Kopyrin I A et al 1979 Optimizing the content of lime, magnesia, and alumina in blast-furnace slags *Metallurgist* **23**:8 536-39

[2] Fedchenko V M and Shparber L Y 1986 Slag regime of a blast furnace *Metallurgist* **30** 351-53

[3] Shiau J S, Liu S H. And Ho C K 2012 Effect of Magnesium and Aluminum Oxides on Fluidity of Final Blast Furnace Slag and Its Application *Mater. Trans., JIM* **53** 1449-55

IOP Conf. Series: Materials Science and Engineering **625** (2019) 012009 doi:10.1088/1757-899X/625/1/012009

[4] Yao L, Ren S, Wang X, Liu Q, Dong L, Yang J and Liu 2016 Effect of Al_2O_3, MgO, and CaO/SiO_2 on Viscosity of High Alumina Blast Furnace Slag *Steel research int.*, **87** 241-49

[5] Zhang K, Wu S, Huang W, Liu X, Zhu J and Du K 2015 Effect of MgO on Emergence of Blast Furnace Primary Slag with Comprehensive Furnace Burden *In: Jiang T. et al. (eds) 6th International Symposium on High-Temperature Metallurgical Processing* Springer, Cham

[6] Nakamoto M, Tanaka T, Lee J and Usui T 2004 Evaluation of viscosity of molten SiO_2–CaO–MgO– Al2O3 slags in blast furnace operation *ISIJ Int.* **44** 2115–19

[7] Panigrahy S C, Rigaud M A J and Dilewijns J 1985 The effect of dolomite addition on the properties of sinters produced from a high aluminous iron ore *Steel Res. Int.* **56** 35–41

[8] Ovchinnikova E V, Gorbunov V B, Shapovalov A N, Maistrenko N A and Bersenev I S 2018 Magnesia Sinter with Flux Based on Magnesium Silicate *Steel in Translation* **48** 34-38

[9] Raygan Sh, Abdizadeh H, Dabbagh A and Pourabdoli M 2013 Influence of talc additive on cold strength and reducibility of iron ore sinters compared to bentonite Ironmaking Steelmaking **36** 273-78

[10] Umadevi T, Nelson K, Mahapatra P C, Prabhu M and Ranjan M 2013 Influence of magnesia on iron ore sinter properties and productivity *Ironmaking Steelmaking* **36** 515-20

[11] Yadav U S, Pandey B D, Das B K and Jena D N 2013 Influence of magnesia on sintering characteristics of iron ore *Ironmaking Steelmaking* **29** 91-95

[12] Shapovalov A N, Ovchinnikova E Vand Maistrenko N A 2015 Improving the Preparation of the Charge Used for Sintering at Ural Steel *Metallurgist* **59** 204-11

[13] Li T, Sun C, Liu X, Song S and Wang Q 2018 The effects of MgO and Al2O3 behaviours on softening–melting properties of high basicity sinter *Ironmaking Steelmaking* **45** 755-63

[14] Ovchinnikova E V, Gorbunov V B, Shapovalov A N, Pisarev S A and Durov L N 2016 Comparative study of the South Urals magnesium materials behavior at the sintering process temperature *Izvestiya vuzov. Chernaya metallurgiya. Trans. Iron and steel* **11** 814-20

[15] Shapovalov A N, Dema R R, Nefedyev S P, Solomonov C N and Yordanova R M 2017 Application of new surfactants sintering agglomerate *Journal of Chemical Technology and Metallurgy* **52** 1008-16

[16] Ganin D R, Druzhkov V G, Panychev A A, Shapovalov A N, Shevchenko E A 2017 Investigation of the effect of additives serpentitomagnezit Khalilovo deposit on the performance of the sintering process at JSC "Ural Steel" *Vestnik of Nosov Magnitogorsk State Technical University* **15** 20-26

Strength and thermal properties of unfired four-hole hollow bricks manufactured from a mixture of cement, low-calcium fly ash and blended fine aggregates

Trong-Phuoc Huynh[1], Trong-Chuc Nguyen[2], Ngoc-Duy Do[3], Chao-Lung Hwang[4], and Le Anh-Tuan Bui[3]

[1] Department of Rural Technology, College of Rural Development, Can Tho University, Campus II, 3/2 St., Ninh Kieu Dist., Can Tho City 900000, Vietnam
[2] Institute of Techniques for Special Engineering, Le Quy Don Technical University, No. 236, Hoang Quoc Viet St., Bac Tu Liem Dist., Ha Noi City, Vietnam
[3] Department of Civil Engineering, College of Engineering Technology, Can Tho University, Campus II, 3/2 Street, Ninh Kieu District, Can Tho City 900000, Vietnam
[4] Department of Civil and Construction Engineering, National Taiwan University of Science and Technology, No. 43, Keelung Rd., Da'an Dist., Taipei City 10607, Taiwan

E-mail: mikehwang@mail.ntust.edu.tw

Abstract. This study develops a new type of eco-friendly building brick so-called unfired four-hole hollow brick (UFHB) by using a mixture of ordinary Portland cement (OPC), low-calcium fly ash (FA), and different proportions of crushed sand (CS) and river sand (RS). A hydraulic-static pressure was applied to form the brick samples. The effect of various CS-RS blends on the mechanical strength of the UFHB samples was studied. In addition, the numerical method was applied to simulate the heat transfer process through the brick wall. The obtained results show that the UFHB mixture containing a mixture of 10% RS and 90% CS as blended fine aggregate registered the highest strength value in comparison with other UFHB mixtures. Moreover, utilization of the UFHB was highly effective in term of heat insulation as compared to the conventional brick.

1. Introduction

The rapid development of construction industry requires a variety of construction materials in a large quantity, especially the demand of building brick. The heat transferring is one of the factors that is mostly considered by both the designer and the user during either the construction process or the use of any building. The wall of a building plays an important role in insulating between the internal and external sides and its primary function is to cover and protect the building from outside factors such as wind, heat, etc. However, the characteristics of the wall rely heavily on the properties of brick inside it. These bricks not only sustain the wall's loading but also supply the great insulation for the building. Recently, more attention has been paid to unfired materials with good physical properties that can fulfill all requirements of strength and serviceability for thermal transmittance. Several scientific publications have been adverted to fire resistance and thermal behavior of a brick or a brick wall, e.g. Bondi and Stefanizzi [1] reported on the hygrothermal performance of hollow bricks; Al Nahhas et al. [2] conducted an experimental and thermal modeling on the resistance to fire of walls constituted by hollow blocks, with the results highlighted the phase change effect under high temperature; Sala et al. [3] performed both the laboratory experiments and numerical analysis on static and dynamic thermal characterization of a hollow brick wall. Besides, the production of conventional brick has been facing with more and more challenges, especially the shortage of natural resources, environmental pollution, and energy consumption. Hence, the development of unfired building brick to replace the conventional one is strongly encouraged. Follow this trend, this study develops unfired four-hole hollow brick (UFHB), a new type of eco-friendly unfired building brick, by using small amounts of OPC and FA with a large proportion of CS and RS. The effect of various CS-RS blends on the compressive strength of the UFHB was investigated. The heat transfer process through the brick wall was also studied.

2. Materials and experimental details

2.1. Materials

Type-PCB40 OPC and low-calcium FA, with characteristics as given in Tables 1 and 2, were used as binder materials, whereas crushed sand (CS) and natural river sand (RS) were used as fine aggregates for this experimental work. Both CS and RS had the respective densities and water absorption (WA) rates of 2.47 and 4.51% and 2.55 and 1.72%. In order to study the effect of aggregate gradation on mechanical properties of bricks, blended CS-RS mixtures as displayed in Figure 1 were especially used. It can be seen that the existence of RS with small particle size modified and reformed the gradation curve and fineness modulus (FM) of the original aggregate.

Table 1. Physical properties of powder materials.

Items	OPC	FA
Specific gravity	3.15	2.29
Mean particle size (μm)	19.1	21.5
Specific surface area (m^2/g)	0.78	0.66
Strength activity index at 28 days (SAI, %)	100	86.5

Table 2. Chemical compositions of powder materials.

Materials	Chemical compositions (wt.%)							
	SiO_2	Al_2O_3	Fe_2O_3	CaO	MgO	SO_3	K_2O	Na_2O
OPC	20.04	4.24	3.12	62.43	4.17	2.97	0.43	0.33
FA	64.01	22.14	5.64	2.75	0.92	0.61	1.36	0.85

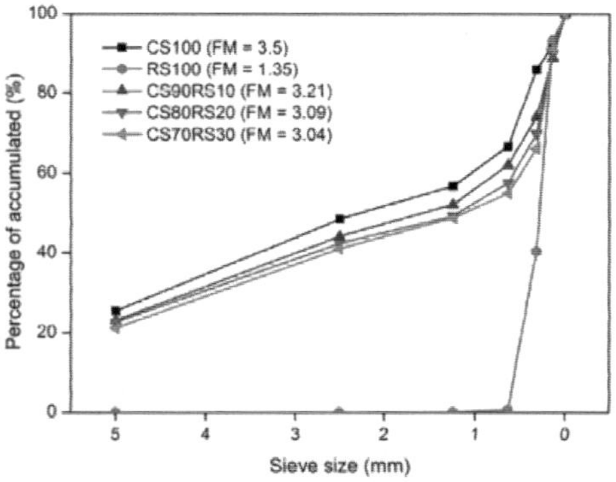

Figure 1. Gradation curves of fine aggregate.

Figure 2. UFHB samples.

2.2. Mix proportioning

In this study, the influence of fine aggregate gradation on the mechanical strength of the UFHB was assessed by using various CS-RS blends. Different percentages of CS (100%, 90%, 80%, and 70%) and RS (0%, 10%, 20%, and 30%) were mixed together to prepare the blended aggregate before making the UFHB samples. A mixture of OPC and FA was used as a binder material at a fixed content of 25% by weight of total mixture. A constant water-to-binder (w/b) ratio of 0.2 was applied for all brick mixtures. The mixture proportions for the preparation of the UFHB samples are given in Table 3.

Table 3. Material proportions for preparing brick samples.

Ingredients (kg/m^3)	UFHB mixtures			
	F15R00	F15R10	F15R20	F15R30
OPC	268.6	268.8	269.0	269.2
FA	335.8	336.0	336.3	336.5
CS	1634.2	1411.3	1188.1	964.6
RS	0.0	224.0	448.3	673.0
Water	120.9	121.0	121.1	121.1

2.3. The preparation of brick samples and experimental methods

According to the designed proportions as presented in Table 3, the 80×80×180 mm UFHB samples were manufactured at a local brick production factory. Herein, a brick's automated production line was used to produce the UFHB samples as shown in Figure 2. It is noted that a hydraulic pressure of about 4.7 MPa was applied to form the samples. The finished bricks were placed in open air for 24h then immersed in water for a haft hour. After that, the bricks were covered by a thin plastic film and stored in the humidity-curing room for 28 days. To evaluate the mechanical strength of the UFHB, the compressive strength test was carried out in accordance with TCVN 6477-2016 [4]. Moreover, the thermal behavior of the produced bricks was evaluated by the simulation of heat transfer across the brick wall.

2.4. Simulation of heat transfer across the brick wall

The basic theory of heat transfer can be presented through equation (1) [5]:

$$\frac{\partial T}{\partial t} = \alpha_T \nabla^2 T \tag{1}$$

Where: $\alpha_T = \lambda / (c.\rho)$ - diffusivity, m^2/s;

c - specific heat, kJ/(kg.^{0}C);

ρ - density, kg/m^3;

λ - thermal conductivity;

$\nabla^2 T = div(grad\ T)$ – Laplace temperature operator.

Applying boundary conditions [6] to solve equation (1):

$$T = T_{outer\ envir.} \text{ and } -\lambda \frac{\partial T}{\partial n} = h(T_s - T_{inner\ room}) \tag{2}$$

Where: $T_{outer\ envir.} = 38^0C$ - the highest average ambient temperature in Southern Vietnam [7];

n - the outward direction normal to the surface;

h - heat transfer coefficient, (W/m^2.^{0}C);

T_s - temperatures at the boundary nodal points of bick wall, ^{0}C;

$T_{inner\ room} = 22^0C$ - average temperature inside the room in use condition.

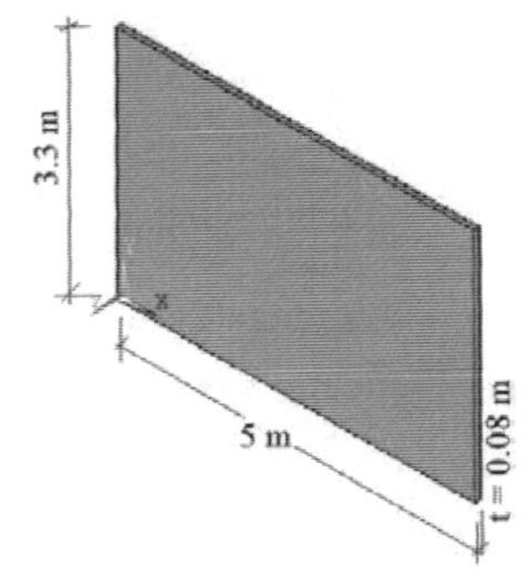

Figure 3. Dimensions of brick wall model.

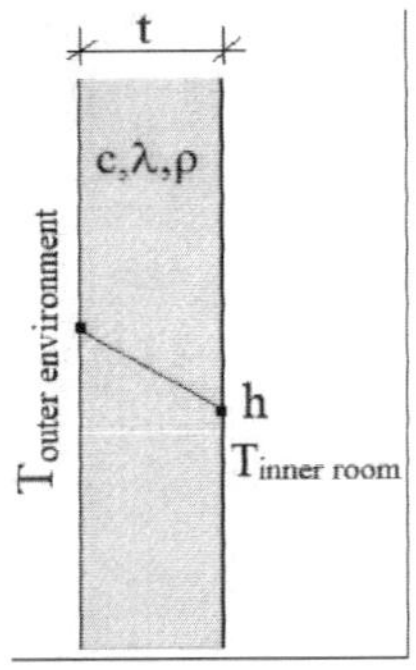

Figure 4. The heat transfer through the brick wall model.

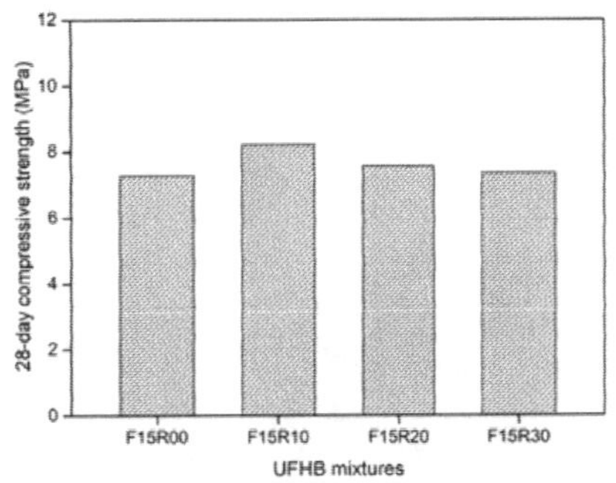

Figure 5. The 28-day compressive strength of UFHB samples.

The effect of convection in the air cavities of the brick holes compared to the air of the environment was shown to be negligible since the air (fluid) was almost still. Therefore, the convection in the air cavities of the brick holes was excluded from the final model. By the way, the computational analysis was simplified and acceptable [8]. In this study, modeling of the brick wall that measured 5.0×3.3×0.08 m (Figure 3) under the influence of environmental temperature in Southern Vietnam was performed. For comparison purpose, it is assumed that two different types of brick, including UFHB and conventional reference brick (CRB), were used to construct the wall. The thermo-physical properties of these brick wall layers and the heat transfer model based on the finite element method (Ansys APDL) are shown in Table 4 and Figure 4, respectively.

Table 4. Thermo-physical properties of brick wall layers.

Thermo-physical properties	Values [9, 10]	
	UFHB	CRB
Thermal conductivity, λ [W/(m·^{0}C)]	0.78	1.25
Density, ρ (kg/m^3)	1900	2090
Specific heat capacity, C [J/(kg·^{0}C)]	1000	920
Convection coefficient, W/(m^2.^{0}C)	25	25

3. Results and discussion

3.1. Compressive strength

The variation in the compressive strength of the UFHB samples with different CS-RS proportions is shown in Figure 5. As the test result, the control mixture comprising 100% CS (F15R00) registered the lowest compressive strength value of 7.27 MPa. It is found that the use of blended CS-RS aggregates resulted in an increase in brick's strength as compared to the control sample. The UFHB samples containing 10% RS in combination with 90% CS had the highest compressive strength value of 8.21 MPa. However, the compressive strength was found to gradually decreasing with the increasing of RS content from 10% to 20% and 30%. In fact, the strength values of the UFHB samples with 20% and 30% RS contents were 7.56 MPa and 7.34 MPa, respectively. This increase in brick' strength was attributable to the filler effect of using RS, which contributed to refine the pores size and thus reduce the void volume within the brick structure. Another possible reason for the increasing strength of the bricks may be the formation of calcite within the cement-sand pore structure [11].

3.2. Thermal behavior of bricks

The temperature distribution during thermal flux through the two types of the brick wall as mentioned previously was analyzed using the famous Ansys APDL computer software, with the outcomes are displayed in Figures 6 and 7. It can be seen clearly that the temperature distribution through the thickness of both the CRB and UFHB walls was linear. The temperatures of about 26.0^{0}C and 27.6^{0}C were recorded at the inside surface of the UFHB and CRB walls, respectively. Thus, the UFHB was found to be more effective than the CRB under the same climate condition. Calculation of heat distribution across the brick wall is very useful to both the designer and the user in order to provide sufficient insulation solutions for each specific climate condition.

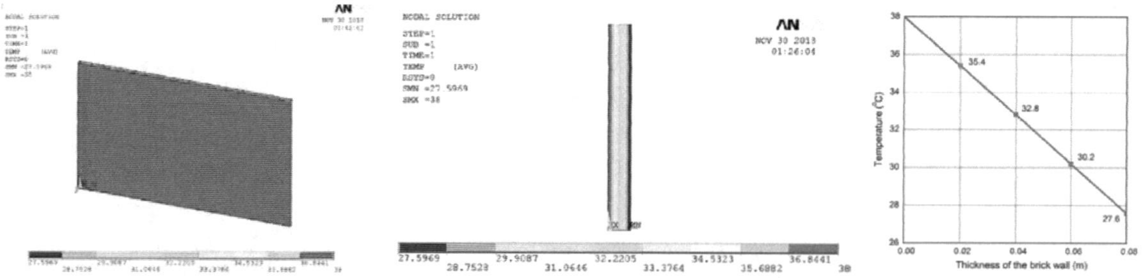

Figure 6. Temperature distribution on the CRB wall.

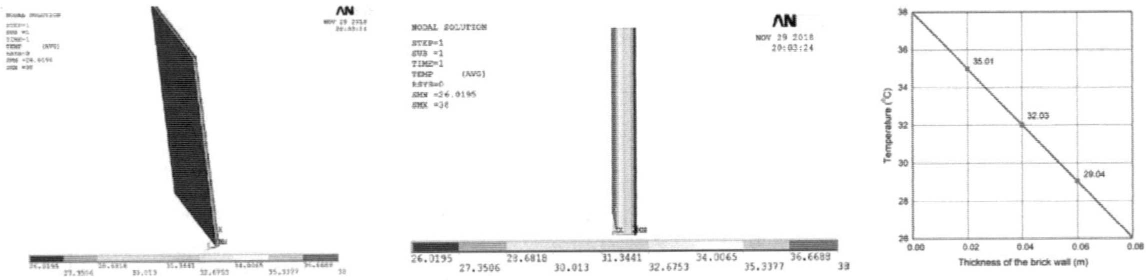

Figure 7. Temperature distribution on the UFHB wall.

4. Conclusions

The following conclusions can be drawn according to the simulation and experimental results:

1. The compressive strength of the UFHB samples was enhanced with an appropriate proportion of RS. This study found that the maximum brick's strength value of 8.21 MPa was achieved by using 10% RS- 90% CS blend and the strength value was declined at further RS addition. In real practice, the compressive strength of the brick is much dominant in comparison with other brick's properties.
2. Analyzing temperature distribution across the brick wall indicates that the use of the UFHB was more effective than using conventional brick in heat insulation aspect. The simulation results could be considered as good reference information in case of applying this type of brick in construction activities.
3. The findings of the present study demonstrate a high potential of using OPC, low-calcium FA, and CS-RS blends in the production of UFHB and further encourage the application of bricks in this type in the construction industry.

Acknowledgments

Special thanks go for Trung Hau Manufacturing Co. Ltd. (Vietnam) for their valuable assistance in the preparation of UFHB samples used for this study.

References

[1] Bondi P and Stefanizzi P Hygro-thermal performance of hollow bricks and current standards 2001 *Energy Build.* **33** 731–736

[2] Al Nahhas F, Ami Saada R, Bonnet G, and Delmotte P Resistance to fire of walls constituted by hollow blocks: Experiments and thermal modeling 2007 *Appl. Therm. Eng.* **27** 258–267

[3] Sala J M, Urresti A, Martín K, Flores I, and Apaolaza A Static and dynamic thermal characterisation of a hollow brick wall: Tests and numerical analysis 2008 *Energy Build.* **40** 1513–1520

[4] TCVN 6477-2016 2016 *Concrete bricks* Ministry of Construction (Vietnam)

[5] Casano G and Piva S Transient heat conduction in a wall exposed to a fire: An analytic approach 2017 *IOP Conf. Series: Journal of Physics: Conf. Series* **796** 012036

[6] Lam V T, Nguyen T C, Boris B, and Pham N A Composition and early-age temperature regime in massive concrete foundation 2018 *MATEC Web of Conferences* **196** 04017

[7] IMHEN and UNDA 2015 *Vietnam's special report on disaster risk management and extreme events in order to promote adaptation to climate change* The Environmental Resources and Maps of Vietnam Publisher (in Vietnamese).

[8] Lacarrière B, Lartigueb B, and Monchoux F Numerical study of heat transfer in a wall of vertically perforated bricks: Influence of assembly method 2003 *Energy Build.* **35** 229–237

[9] Kanellopoulos G, Koutsomarkos V G, Kontoleon K J, and Georgiadis-Filikas K Numerical analysis and modelling of heat transfer processes through perforated clay brick masonry walls 2017 *Procedia Environ. Sci.* **38** 492–499

[10] Tamene Y, Abboudi S, and Bougriou C Study of heat and moisture diffusion through a wall exposed to solar heat flux 2011 *J. Eng. Sci. Technol.* **6** 429–444

[11] Thomas B S Green concrete partially comprised of rice husk ash as a supplementary cementitious material – A comprehensive review 2018 *Renew. Sust. Energ. Rev.* **82** 3913–3923

IOP Conf. Series: Materials Science and Engineering **625** (2019) 012011 doi:10.1088/1757-899X/625/1/012011

The enhancement of high dielectric properties for pure PVDF-HFP by using the suitable stirred temperature and time

A Salea[1], K Jehlaeh[1] and C Putson[1,*]

[1] Materials physics laboratory, Physics department, Faculty of Science, Prince of Songkla University, Songkhla, 90112, Thailand

*Email: chatchai.p@psu.ac.th

Abstract. A novel light-weight electronic device from the dielectric polymer is so remarkable as the actuators, sensors, energy storage device and energy conversion. For this work, the thin film electroactive polymer is fabricated by casting technique which PVDF-HFP dissolve in *N,N*-dimethylformamide (DMF). The three stirred temperatures (30, 55 and 80 °C) and two stirred times (6 and 12 h) are completely considering. The highest polarization as the β phase is also considered for dielectric properties improvement. The FTIR, LCR and DSC equipment are used to measure the electroactive improvement. As a result, the 12 h stirred time with 30 °C stirred temperature is the best condition. The 0.25 g/ml concentration is also the suitable condition. The better dielectric properties are the requisite condition for a novel energy storage device in the future.

1. Introduction

The flexibility, light-weight, easy to process and high breakdown strength of dielectric polymer have to be the main reason to apply in such the novel applications as the energy storage, electrostriction, pyroelectric and electrocaloric [1, 2]. These advantages of the polymer are so interesting to improve dielectric properties as much as dielectric ceramic materials. Polyvinylidien fluoride-hexafluoropropylene (PVDF-HFP) is a semi-crystalline polymer which is one of high dielectric properties. The improvement of dielectric properties has to controllable PVDF-HFP chain structure to be the highest net-dipolemoment as called β phase. The β phases are -[CH_2-CF_2]- structure that arranged to the opposite site between two C-H and two C-F, which is bonding up and down perpendicular to carbon backbone causing to produce the highest net-dipolemoment per unit cell compared with another phases such as α, γ and δ [3, 4]. The stretching and poling techniques are able to response the β phases to enhance including dielectric properties [5]. The controllable PVDF-HFP structure is able to improve β phases according to 511, 837, 1234 and 1275 cm^{-1} vibrational peak in FTIR [6]. Several reported papers tried to improve dielectric properties by using composites powder for dispersion in the polymer matrix but they lack observed the suitable stirred temperature and stirred time especially in the pure PVDF-HFP polymer. The suitable stirred temperature and stirred time might able to create the higher β phase transformation causing to enlargeable dielectric properties.

2. Materials and method

Polyvinylidien fluoride-hexafluoropropylene (P(VDF-HFP) powder, purchased from Solvay Solexis, Belgium) is dissolved in 99% purity *N,N*-dimethylformamide (DMF, purchased from RCI Labscan Limited, Thailand). And it is then stirred 12 h in different temperatures (30, 55 and 80 °C) to be homogeneous solution before cast on the smooth class plate. They are directly dried in the oven at 80 °C for 24 h to become 100 μm thin film as showed in **Figure 1 (a)**. Afterward, their electroactive β phase and dielectric properties are analysed by using FTIR technique and LCR meter in frequency from 1 to 10^5 Hz followed by using the equation (1) and (2), respectively. Moreover, it are suddenly confirmed the melting temperature by using DSC technique. In addition, the best condition is altered the stirred time to be 6 h (from 12 h). Finally, the best condition is observed the surface morphology by using SEM with EDX for observing the atomic distribution.

3. Results and discussion

3.1. The electroactive β phases transformation

Fourier Transform Infrared Spectra (FTIR), (Vertex70, Bruker, Germany), is scientific equipment that is able to classify PVDF-HFP structure. **Figure 1 (b-c)** represented to the electroactive β phase including with α and γ (figure 2 a). The transformation of β phases are confirmed in the growing β phase intensity by using Lambert-Beer law the equation (1) which is a ratio between α and β phase at the wavenumber of 767 and 837 cm^{-1} respectively.

$$F(\beta) = \frac{A_\beta}{\frac{K_\beta}{K_\alpha}A_\alpha + A_\beta} \quad (1)$$

Where, A_α is the absorbance at 767 cm^{-1} and A_β is the absorbance at 837 cm^{-1} and K_α (6.1 x 10^4 cm^2 mol^{-1}) and K_β (7.7 x 10^4 cm^2 mol^{-1}) are the absorption coefficients at respective wavenumber [7]. The absorbance of wavenumber around 830 - 840 cm^{-1} is able to obviously seem that the β phase decrease when they are increased the stirred temperature to 55 and 80 °C, respectively. The highest β phase is showed at 30 °C. The β phase is calculated and plotted in different stirred temperature as showed in figure 2 (b). The higher temperature might be the disturbing the β phase conformation to be another phases as α or γ.

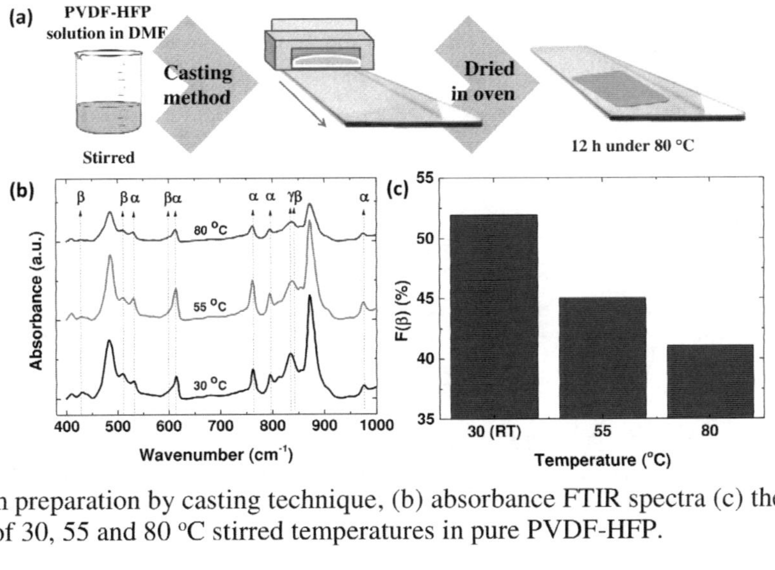

Figure 1 (a) film preparation by casting technique, (b) absorbance FTIR spectra (c) their absorbance β phase intensity of 30, 55 and 80 °C stirred temperatures in pure PVDF-HFP.

3.2. Dielectric properties and thermoanalytical analysis

The LCR meter (IM 3533 HIOKI) is dielectric measurement equipment to measure such dielectric properties as dielectric constant (ε_r) and dielectric loss (ε_r'') in various frequencies (1 - 10^5 Hz). The samples are applied 1 V ac voltage pass through the sample by two electrodes. And then they are measured surface charge (C, in F) and calculated to be ε_r as showed an Equation (2).

$$\varepsilon_r = Cd/\varepsilon_0 A \quad (2)$$

Where d is sample thickness (m), A is electrode area (m^2), and ε_0 is free space permittivity with value 8.853 x 10^{-12} F m^{-1}[8]. **Figure 2 (a-b)** showed the dielectric properties. The dielectric constant is decreased with increasing the frequency in any conditions as showed in figure 3 (a) which can be explained by Maxwell-Wagner-Sillars (MWS) interfacial polarization [9]. Moreover, as dielectric constant with different temperatures, the dielectric constant decreases when the temperature is increased because of the disturbing β phase from heat following FTIR result. In addition, the dielectric loss in figure 3 (b) is insignificantly different that the loss at the lowest frequency (1 Hz) produces unsteady transient because the lower frequency make easy to rearrange chain polarization or unsteady loss and it might fluctuate polarization.

Differential Scanning Calorimetry (DSC), Simultaneous Thermal Analyser, STA8000, Perkin Elmer, USA, is thermoanalytical technique to measure the absorbed heat energy of the sample from 30 °C (room temperature, RT) to 200 °C in the air atmosphere as showed **Figure 2 (c).** The increasing stirred temperature is able to ship the melting temperature (T_m) referred to the changing of crystallinity to be lower temperature from 160.68 to 159.55 and 159.07 °C for 30, 55 and 80 °C, respectively. This result can also confirm the previous data that the hotter stirred temperature might able to destroy net polarization from the β phase to be another.

The stirred time of the best solution (RT) is decreased from 120 h to 6 h to find the minimum time for the suitable stirred time with measure dielectric properties as showed **Figure 2 (d-e)**. It can seem that the dielectric constant increase when the stirred time is increase from 6 to 12 h because it have time to rearrange and increase the β phase to be greater as well as their dielectric loss.

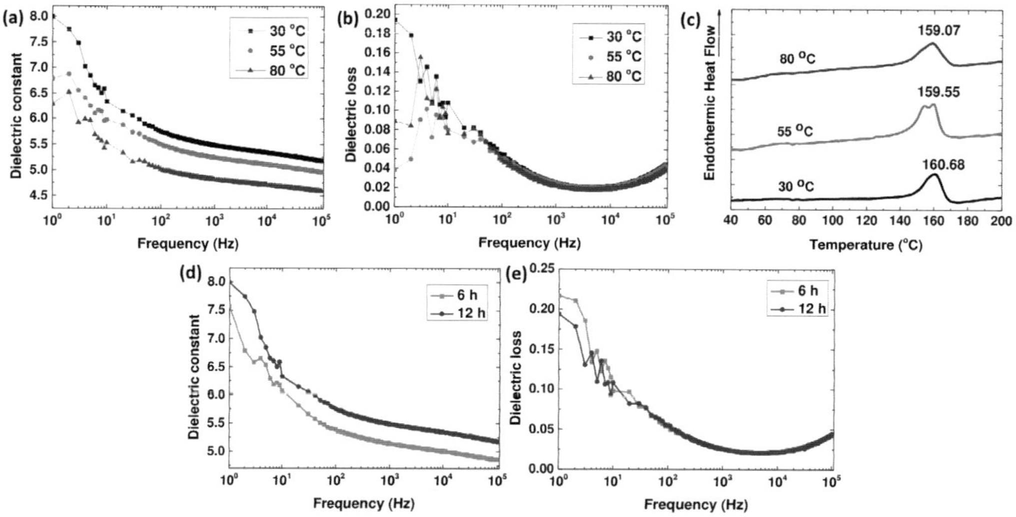

Figure 2 Dielectric constant (a), dielectric loss (b) and DSC-thermograph (c) of 30, 55 and 80 °C three stirred temperatures (d) dielectric constant and (f) dielectric loss of 6 and 12 h stirred times in pure PVDF-HFP

3.3. Surface morphology, atomic distribution and the thickness with various concentrations.

Scanning Electron Microscopy with Energy Dispersive X-ray spectroscopy (SEM-EDX), FEI, Quanta 400, SEM-Quanta, is electron microscope that not only produces the image by electron scattering but also detects several atomic elements which distribute on the sample surface as showed in **Figure 3 (a-d)**. The surface morphology of the sample is also slightly rough of porous which might come from the evaporated solvent as showed in figure 4(a). The carbon (C) and fluorine (F) atoms are detected for PVDF-HFP structure which distributed on that surface as showed in figure 4 (b-c). The structure of PVDF-HFP is shown in figure 4 (d) which related to the ratio of both two elements.

The PVDF-HFP with DMF is proportionated to 0.20, 0.25 and 0.30 g/ml as called concentration with measure thickness in various blade sizes (1.2, 1.4 and 1.6 mm) and figure out the relationship as showed in **Figure 3 (e)**. It can seem that the thickness size of samples is increased with increasing concentration and blade sized because DMF are evaporated due to heat. The 0.25 g/ml concentration, 1.2 blade size and 100 μm thickness size is suitable for this work due to dissolvable and savable amount of DMF.

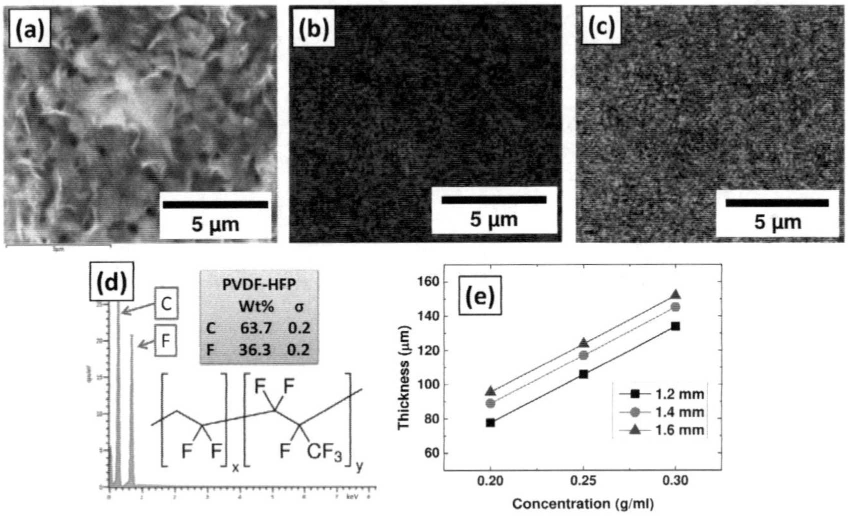

Figure 3 Surface morphology on pure PVDF-HFP; SEM image (a), with EDX signals of C (b) and F (c) elements, the percent atomic weight with PVDF-HFP structure, (d) relationship between thickness sizes with three concentrations in different blade sizes.

4. Conclusion

The 0.25 g/ml of PVDF-HFP in DMF is the suitable concentration for this work. It is cast on the smooth glass to adjust the thickness for characterization. The highest polarization as β phase is a critical dielectric properties term. The three stirred temperatures and two stirred time are considered. FTIR, Dielectric properties including with DSC are analysed. The best condition in this work is 12 h stirred time and 30 °C stirred temperature. The highest dielectric properties are applicable to be energy storage devices applications including a novel light-weight battery.

Acknowledgments

We would like to thanks the Thailand Center of Excellence in Physics (ThEP-61-PIP-PSU3), Science Achievement Scholarship of Thailand (SAST), including Physics Department, Faculty of Science, and Prince of Songkla University, Songkhla, Thailand.

References

[1] Lallart M, Thetpraphi K and Capsal J-F. 2018 *Phys. Lett. A* **382** 449-54
[2] Yang L, Qian XS, Koo C, Hou Y, Zhang T, Zhou Y, Lin MR, Qiu JH and Zhang QM. 2016 *Nano Energy* **22** 461-7
[3] Chen S, Yao K, Tay FEH and Liow CL. 2007 *J. Appl. Phys.* **102** 104108
[4] Martins P, Costa CM, Benelmekki M, Botelho G and Lanceros-Mendez S. 2012 *CrystEngComm* **14** 2807-11
[5] Kepler RG and Anderson RA. 1978 *J. Appl. Phys.* **49** 1232-5
[6] Parangusan H, Ponnamma D and Al-Maadeed MAA. 2018 *Sci. Rep.* **8** 754
[7] Thakur P, Kool A, Bagchi B, Das S and Nandy P. 2015 *Phys. Chem. Chem. Phys.* **17** 1368-78
[8] Zhong K and Li B 2017 *Polymer Nanocomposites for Dielectrics* (Singapore: Pan Stanford Publishing Pte. Ltd) p 208
[9] Tsangaris GM, Psarras GC and Kouloumbi N. 1998 *J. Mater. Sci.* **33** 2027-37

IOP Conf. Series: Materials Science and Engineering **625** (2019) 012012 doi:10.1088/1757-899X/625/1/012012

Microstructures of nanoporous-Si composite anodes in sulfide-based all-solid-state lithium-ion batteries

R Okuno[1], M Yamamoto[1,2], A Kato[2], Y Terauchi[2] and M Takahashi[1,2*]

[1] Nara Institute of Science and Technology, 8916-5 Takayama-cho, Ikoma, Nara 630-0192, Japan

[2] Osaka Research Institute of Industrial Science and Technology, 1-6-50 Morinomiya, Joto, Osaka 536-8553, Japan

[*] E-mail: masataka@omtri.or.jp

Abstract. Nanoporous-Si composite anodes with different dispersibility were prepared to compare the electrochemical performances of sulfide-based all-solid-state lithium-ion batteries. Ball-milled anode with higher dispersibility exhibited the capacity retention of 80% at 150th cycle. The structure and composition were accurately analyzed using field emission scanning electron microscope with energy dispersive x-ray spectrometer. The both of the pores in Si particles and the elastic deformation of solid electrolyte effectively relieved the stress derived from the volume change of highly dispersed nanoporous Si, resulting in the high capacity retention.

1. Introduction

All-solid-state lithium-ion batteries (LIBs) with high safety and reliability are the dominant power source for electric and hybrid vehicles. To meet the energy demands of near-future automobile technology, the anode materials should possess the enhanced storage capacity, high energy density, and long cycle life. Though Si is the most attractive choice because of its high theoretical capacity (4200 mAh g^{-1}), there are few applications for the all-solid-state LIBs [1,2]. In contrast, significant problems prevent the practical use of Si in the LIBs using liquid electrolytes. The large volumetric fluctuation (>300%) in lithiation/delithiation causes the mechanical fracture of Si-based anodes and repeats the formation of surface electrolyte interface (SEI). These behavior result in the poor cycle life, drastic irreversible capacity loss, and low coulombic efficiency (CE). To overcome these issues, nanosizing (e.g., nanowire, nanotube, hollow nanosphere) is effective to relieve the structural stress [3,4]. Despite the superior performance, these promising candidates have disadvantages such as the high manufacturing costs and handling difficulty.

Our interest focuses on porous particles, because the pores are expected to act as the buffer region for volume change. Very recently, we have found that all-solid-state LIB with nanoporous-Si composite anode exhibited high capacity retention [5]. In general, the performance of LIBs strongly depends on the microstructural features of electrodes. Therefore, it is important to optimize the systematic design of porosity, dispersibility, particle size and shape. In the present study, the nanoporous-Si composite anodes with different dispersibility were prepared by ball milling and mortar milling to elucidate the cause of the observed stable cyclability. We report here the results of structural, compositional, and electrochemical studies on the sulfide-based all-solid-state LIBs with nanoporous-Si composite anodes.

2. Experimental procedures

2.1. Preparation

Nanoporous-Si particles were prepared through air-oxidation demagnesiation of Mg_2Si using Qian's method [6]. Mg_2Si (1000 mg) and Mg (30 mg) were mechanically milled at 1800 rpm for 15 min. The mixture was oxidized in air at 700 °C for 12 hrs. All of the following procedures were performed under dry Ar atmosphere. Si particles were collected by centrifugation after washing the oxidation product with HCl, followed by deionized water and ethanol three and two times, respectively. Solid electrolyte (SE), Li_3PS_4 glass, was prepared by mechanical milling of Li_2S and P_2S_5 [7]. Anode composite material was comprised of the nanoporous-Si particles (36wt%), SE powder (55wt%), and acetylene black (AB: 9wt%). We adopted two mixing methods, ball milling and mortar milling, to vary the dispersion degree of Si particles. Mechanical ball milling was conducted by using high speed mixer (CM-1000, EYELA) equipped with ZrO_2 balls (ϕ4 mm and ϕ1 mm) under rotation speed of 1800 rpm for 60 min. Mortar milling was carried out for 10 min.

All-solid-state half-cells were fabricated as follows. In an electric insulation tube with 10 mm diameter, the SE powder (80 mg) and anode composite material (2 mg) were pressed under 330 MPa to make two-layered pellet. As the counter electrode, Li-In foil was attached on the SE side. Finally, the three-layered pellet was compressed at 70 MPa using stainless-steel disks.

2.2. Structural, compositional, and electrochemical measurements

XRD patterns of the nanoporous-Si particles were measured with Rigaku SmartLab using CuKα radiation. To accurately analyze the cross section of the nanoporous-Si composite anodes, the samples were transferred to a polishing machine (IB-19520CCP, JEOL) using vacuum vessel. The surface was carved with 4 keV Ar^+ ion beam, the ion current being 70-80 μA. The structure and composition were evaluated using field emission scanning electron microscope (FE-SEM: JSM-7800F, JEOL) with energy dispersive x-ray spectrometer (EDX).

By a charge/discharge measurement device (BTS-2004, Nagano), the electrochemical tests were conducted in the constant current mode of 0.127 mA cm^{-2} for initial 3 cycles and 0.3 mA cm^{-2} for the following cycles at 30 °C. The cut-off voltages were 0.88 and -0.58 V vs Li-In, corresponding to 1.50 and 0.04 V vs Li^+/Li.

3. Results and discussion

3.1. Structural and compositional characteristics

FE-SEM image of the grain comprising nanoporous-Si particles is shown in figure 1. The disordered nanometer pores were clearly observed on the grain surface. By applying the Scherrer equation to XRD data, the average crystallite size of Si particles was estimated to be about 55 nm. Figure 2 shows EDX mapping images of the nanoporous-Si composite anodes. Si and SE are colored in white and black, respectively. In the ball-milled anode, nm-sized Si particles were highly dispersed in the SE. On the other hand, some Si particles were aggregated in the mortar-milled anode. Since the both anodes contain the same quantity of nanoporous-Si particles, the observed difference is caused by the mixing methods. Namely, the dispersibility in the ball-milled anode is concluded to be higher than that in the mortar-milled anode.

IOP Conf. Series: Materials Science and Engineering **625** (2019) 012012 doi:10.1088/1757-899X/625/1/012012

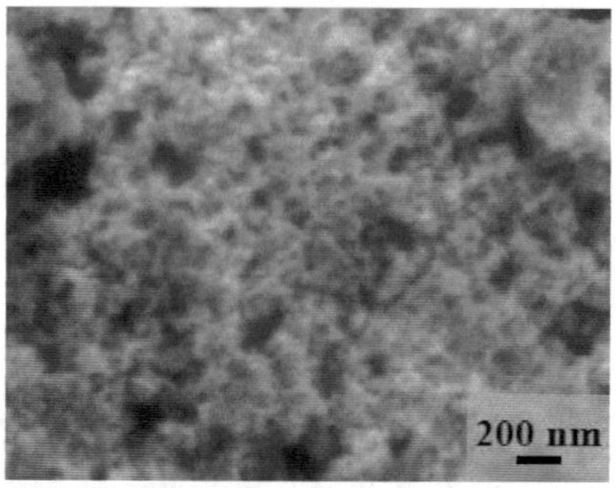

Figure 1. FE-SEM image of nanoporous-Si grain.

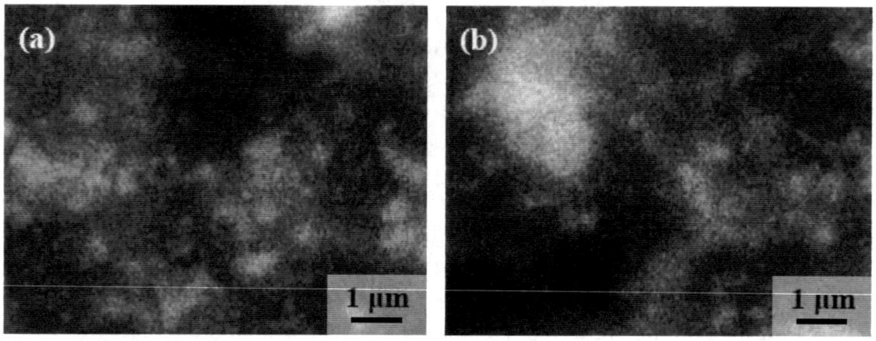

Figure 2. Cross-sectional EDX mapping images of (a) ball-milled and (b) mortar-milled nanoporous-Si composite anodes. Si and SE are colored in white and black, respectively.

3.2. Electrochemical characteristics

Figure 3(a) shows the first charge/discharge curves of the half-cells with ball-milled and mortar-milled nanoporous-Si composite anodes. The former possessed the initial charge and discharge capacities of 2543 and 1240 mAh g^{-1}, respectively, resulting in CE of 49%. In the latter, their values were 3259 and 2300 mAh g^{-1}, corresponding to 71% CE. Figure 3(b) shows the cycle characteristics of ball-milled and mortar-milled anode cells. It is worthy to note that the discharge capacity was maintained in the former, resulting in the capacity retention of 80% at 150th cycle. In contrast, the lower capacity retention, 44% at 150th cycle, was obtained in the latter. To precisely estimate the capacity retentions, they were calculated from the maximum discharge capacities at 15th and 4th cycles for the former and latter, respectively. Previously, non-porous nano-Si composite anode cells have exhibited the poorer electrochemical performances with the discharge capacity less than 1000 mAh g^{-1} at scant 20th cycle [2]. Hence, the retained capacities are extremely high in the nanoporous-Si composite anode cells.

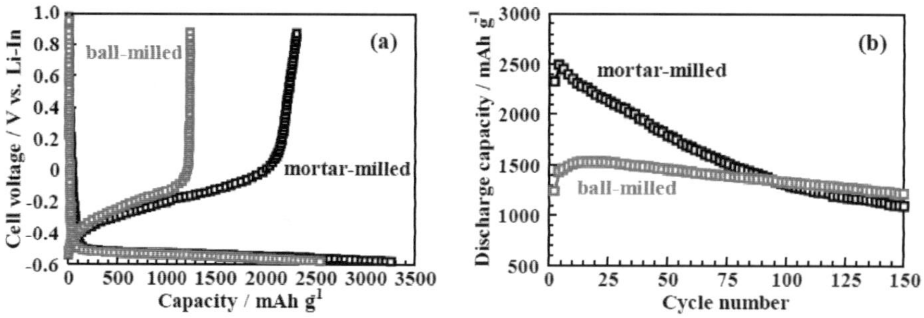

Figure 3. (a) Charge/discharge curves during the first cycle and (b) plot of discharge capacity versus cycle number of sulfide-based all-solid-state half-cells with ball-milled and mortar-milled nanoporous-Si composite anodes.

3.3. Microstructures
To consistently explain the structural, compositional, and electrochemical features described above, we propose the microstructural models for the nanoporous-Si composite anodes. Specifically, the nanoporous-Si particles are highly dispersed in the ball-milled anode (figure 4(a)), while μm-sized Si aggregates exist in the mortar-milled anode (figure 4(b)).

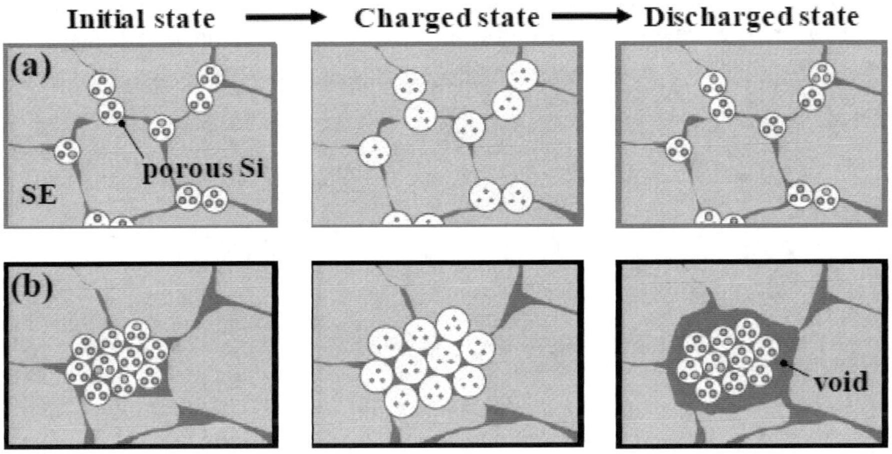

Figure 4. Microstructural models of (a) ball-milled and (b) mortar-milled nanoporous-Si composite anodes.

Since a lot of pores on the surface of Si particles are exposed to SE in the ball-milled anode, SEI should be easily formed at the first cycle, leading to the large amount of irreversible Li consumption. This finding successfully explains that the initial CE is smaller than that of the mortar-milled anode cell. However, SEI is hard to become thick with the increase of charge/discharge cycle. This is because SE possesses high stability and low fluidity in comparison with liquid electrolytes [8].

In regards to the capacity retention, we note the contact state between SE and Si particles/aggregates. Though the volumetric expansion is buffered by the shrinkage of pores, the nanoporous-Si particles slightly become large during lithiation. In the case of highly dispersed Si particles (figure 4(a)), the stress arising from the expanded Si particles should be relieved by the elasticity of surrounding SE. During delithiation, elastically deformed SE returns to the original shape and maintains the contact with Si. In contrast, the μm-sized nanoporous-Si aggregates should give larger stress to the surrounding SE, leading to the plastic deformation of SE (figure 4(b)). Thus, in the delithiation process, voids are formed between the Si aggregates and plastically deformed SE, resulting in the lower capacity retention.

4. Summary
Sulfide-based all-solid-state LIB with ball-milled nanoporous-Si composite anode exhibited the capacity retention of 80% at 150th cycle. Composition mapping analysis indicated that nm-sized Si particles were highly dispersed in the SE-AB matrix. Based on the comparative results with the mortar-milled cell, the highly dispersed nanoporous-Si particles form the intimate contact with SE due to the buffer effect of pores and elastic deformation of surrounding SE. In these respects, we currently make effort to compare the structure and composition between the before and after charging/discharging by high-resolution FE-SEM/EDX system.

Acknowledgments
The authors gratefully acknowledge Visiting Professor Yutaka Fujiwara and Yasuyuki Agari (Nara Inst. Sci. Tech.) for their valuable discussions. This work was partially supported by JSPS KAKENHI Grant Numbers JP16K06787, JP16K05949.

IOP Conf. Series: Materials Science and Engineering **625** (2019) 012012 doi:10.1088/1757-899X/625/1/012012

References

[1] Yamamoto M, Terauchi Y, Sakuda A and Takahashi M 2018 *J. Power Sources* **402** 506

[2] Piper D M, Yersak T A and Lee S-H 2013 *J. Electrochem. Soc.* **160** A77

[3] Wu H and Cui Y 2012 *Nano Today* **7** 414

[4] Liu X H, Zhong L, Huang S, Mao S X, Zhu T and Huang J Y 2012 *ACS Nano* **6** 1522

[5] Okuno R, Yamamoto M, Terauchi Y and Takahashi M 2018 *Energy Procedia* in press

[6] Liang J, Li X, Hou Z, Guo C, Zhu Y and Qian Y 2015 *Chem. Commun.* **51** 7230

[7] Yamamoto M, Takahashi M, Terauchi Y, Kobayashi Y, Ikeda S and Sakuda A 2017 *J. Ceram. Soc. Japan* **125** 391

[8] Fergus J W 2010 *J. Power Sources* **195** 4554

IOP Conf. Series: Materials Science and Engineering **625** (2019) 012013 doi:10.1088/1757-899X/625/1/012013

Mechanical strength of natural rubber filled fly ash

A Dasaesamoh, T Osotchan and K Subannajui

Materials science and engineering program, Faculty of science, Mahidol university, 272 Rama VI Road, Ratchathewi District, Bangkok 10400, Thailand

Email: dabeding@gmail.com

Abstract. In this present work, Fly ash was incorporated into natural rubber to form a natural rubber composite. The natural rubber composite was prepared by solution mixing method at the constant amount of fly ash at 80 phr. The fly ash particle was modified its surface by TESPT silane coupling agent before mixing with natural rubber latex. The surface morphology of natural rubber composites were examined by SEM image. The vulcanization parameters were examined by MDR technique. The results show that modified fly ash affects to natural rubber compound by shorten the vulcanization time and increase maximum torque. The mechanical properties were investigated from stress-strain curve. It reveals that the surface modification of fly ash improves tensile strength of natural rubber by 6 times. The increasing of strength may attribute to the formation of polymer-fillers interaction through TESPT molecules.

1. Introduction

Fly ash, a by-product from power generation industrials, was considered as environmental pollution. The chemical component of fly ash was varied depended on source of materials. In general, it comprise of oxides such as SiO_2, Al_2O_3, Na_2O, K_2O, CaO, and MgO. Many researches has continually studied to minimize these problems by using a fly ash as raw materials for various applications such as car tire[1], geopolymer cement[2], brick[3], mosaic tile[4] by mean of pure fly ash or form in composite polymer. However, less attention has been paid to incorporate fly ash into natural rubber for mechanical properties improvement. The aim of this work was to prepare the polymer composite of fly ash and natural rubber at constant fillers loading of 80 phr. The filler distribution in natural rubber matrix was observed from SEM image. The vulcanize characterization including vulcanization time, rate of vulcanization as well as reversion effect was determined from cure curve. To improve the compatibility, fly ash was modified surface by silane coupling agent and observed the tensile strength from stress-strain curve.

2. Materials and method

2.1. Materials

The fly ash samples were a by-product from biomass electric power generation and kindly supported form Gulf Yala Green, Yala, Thailand. The fly ash particles are in dark green color and flake shape (figure 1). The fly as consists of oxides of K_2O (14.27%), CaO (39.88), SO_3 (2.76%), MgO (5.67%), SiO_2 (4.59%), Fe_2O_3 (0.59%) and Al_2O_3 (0.01%) with loss on ignition (LOI) of 21.34. The summation of SiO_2, Al_2O_3 and Fe_2O_3 was less than 10 %.

IOP Conf. Series: Materials Science and Engineering **625** (2019) 012013 doi:10.1088/1757-899X/625/1/012013

(a) (b)

Figure 1. (a) Physical appearance (b) SEM image of Fly ash

The natural rubber latex with 60 % dry content and vulcanization agents such as wax, ZnO, steric acid, TMTD, Wing stay L and sulphur were purchased from Chemical & Materials co., Ltd, Bangkok, Thailand. Bis[3-(triethoxysilyl)propyl] tetrasulfide (TESPT), silane coupling agent, was obtained from Kisco (T) Ltd, Bangkok, Thailand.

2.2. Composite preparation

The fly ash was first modified with TESPT by drop a 2% of TESPT in a mixture solution of ethanol and deionized water (90:10 %wt.) and vigorously stirred for 2 hr. Then, the TESPT solution was mixed with fly ash at 50 %wt. Finally, the mixing was dried in an oven at 70 °C for 18 h.

Table 1. Vulcanization agents for composite.

Chemicals	phr
Wax	5
ZnO	4
Steric acid	2
TMTD	0.5
Wing stay L	1
Sulfur	2.5

The natural rubber composite was prepared by disperse the fly ash into natural rubber latex and thoroughly mixed for 10 min. The solution was coagulated by 2 % of acetic acid and washed with DI until its pH equal 7. The sample was dried in an oven at 70 °C for 14 h before addition of a vulcanize agents. The vulcanize agents include wax, ZnO, steric acid, TMTD, Wing stay L and sulphur were added into natural rubber composite on two-roll mill machine. All vulcanize agents are in phr unit (part per hundred part of rubber) as shown in table 1. Finally, the composite was formed into sheet through a nap before vulcanization under an electrically heated hydraulic press at 150 °C followed their vulcanization time obtained from MDR results.

2.3. Characterization

The dispersion of fly ash in NR matrix was observed by SEM. The cure characterization was carried out at 150 °C using Moving Die Rheometer (MDR) technique followed with ASTM D5289. The mechanical properties including tensile strength, elongation at break and modulus at 100 % was done using Universal testing machine (INTRON 3366) according to ASTM D412 (Die C).

IOP Conf. Series: Materials Science and Engineering **625** (2019) 012013 doi:10.1088/1757-899X/625/1/012013

3. Results and discussion

The dispersion of fly ash in natural rubber matrix was observed by SEM images as show in figure 2. As can be seen, fly ash shows a good dispersion in natural rubber matrix. It also found some void in both composite polymers, which can future effect to mechanical properties.

The vulcanization parameters of sample were examined as shown in figure 3(a) and the data are summarized in table 2. It reveals that scorch time (t_s2) is shorten when fly ash were incorporated. The shortening of time for starting the crosslinking reaction may due to its pH of fly ash which is base form. The basicity of fly ash may accelerate crosslink reaction[5]. The maximum torque (M_H) of modified fly ash is highest. This implied that the numbers of crosslinks were formed through the silane coupling agent. In addition, the reduction of M_H in fly ash/NR composite may arise from filler agglomerations. The vulcanization time of neat natural rubber, fly ash and modified composite indicated by t_C90 were 3.30, 3.00 and 3.24, respectively. These reveal that incorporation of fly ash whether modified surface or not shorten the time of crosslinkage formation. Moreover, incorporation of fly ash or modified fly does not cause the desulfidic of crosslinking called reversion effect after fully form a network structure. This behaviour preserves the composite materials crosslinkage form over curing.

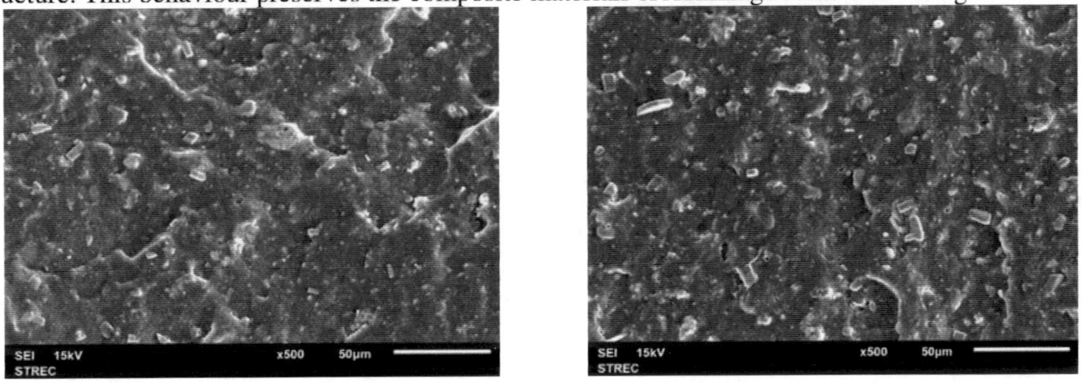

(a) fly ash – NR composite (b) Modified fly ash – NR composite

Figure 2. SEM images of fly ash – NR composite

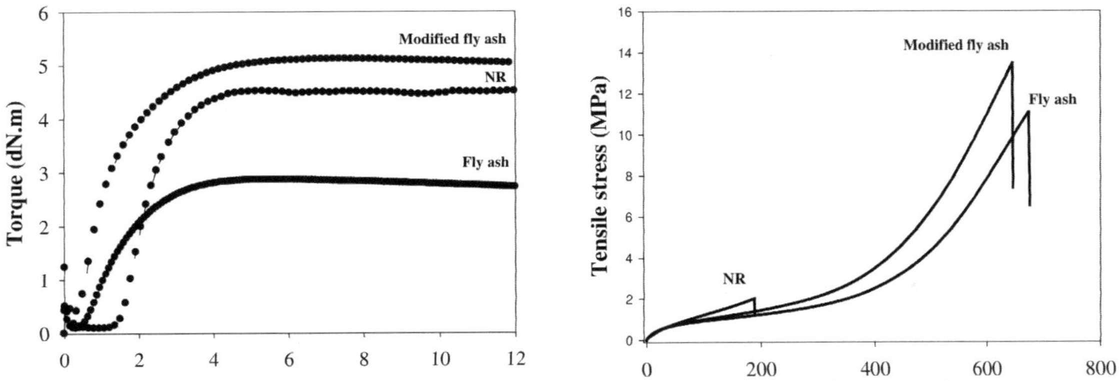

a) b)

Figure 3. a) Vulcanization curves and b) stress-strain curves of neat and natural rubber composites

Table 2. Vulcanization parameters of composite materials.

Compound	M_L (dN.m)	M_H (dN.m)	M_H- M_L (dN.m)	t_s2 (min)	t_C90 (min)
neat natural rubber	0.19	7.60	4.63	2.37	3.30
NR-fly ash	0.15	2.87	2.72	2.13	3.00
NR-modified fly ash	0.62	5.11	4.49	1.06	3.24

Table 3. Mechanical properties of composite materials.

Compound	Modulus at 100 % (MPa)	Tensile strength (MPa)	Elongation at break (%)
neat Natural rubber	1.22	2.1	192
NR-fly ash	0.92	10.9	673
NR-modified fly ash	0.98	13.2	645

The samples were cured at their vulcanization time according to MDR results under the compression and form into 2 mm thickness for mechanical test. The stress-strain curve and listed mechanical properties are shown in figure 3(b) and table 3, respectively. It reveals that the mechanical properties were improved by incorporation of fly ash. Surface modification by silane coupling agent future increases tensile strength. This is due to the silane coupling agent act as bridge molecules connecting natural rubber chain and fly ash results in increase polymer- fillers interaction of composite materials[6]. In addition, elongations at break of natural rubber composite were higher than that of neat natural rubber. However, when fly ash was modified surface, the elongation at break decreased slightly. This behaviour may owing to the enhanced interface interaction of silane coupling agent.

4. Conclusions

1. The inclusion of fly ash into natural rubber effects the vulcanization time of fly ash /natural rubber composite.

2. The modification of surface by silane coupling agent improves mechanical properties by increasing the degree of fillers – polymer interactions.

Acknowledgments

The authors would like to thank office of the Higher education commission of Thailand for financial supporting.

References

[1] Ren X, Sancaktar E 2019 *J.Cleaner Production* **206** 374.
[2] Zhuang X.Y, Chen L, Komarneni S, Zhou C.H, Tong D.S, Yang H.M, Yu W.H, Wang H 2016 *J. Cleaner Production* **125** 253.
[3] Naganathan S, Mohamed A.Y.O, Mustapha K.N 2015 *J.Construction and Building Materials* **96** 576.
[4] Luo Y, Zheng S, Ma S, Liu C, Wang X 2017 *Ceramics International* **43** 11953.
[5] Ooi Z.X, Ismail H, Abu Bakar A 2013 *Polymer Testing* **32** 38.
[6] Thongsang S, Sombatsompop N 2006 *Polymer Composites* **27** 30.

IOP Conf. Series: Materials Science and Engineering **625** (2019) 012014 doi:10.1088/1757-899X/625/1/012014

Furfuryl alcohol-derived carbon monoliths for CO2 capture: adsorption isotherm and kinetic study

J Singh[1], S Basu[1] and H Bhunia[2]

[1] School of Chemistry and Biochemistry, Thapar Institute of Engineering and Technology, Patiala 147004, Punjab, India

[2] Department of Chemical Engineering, Thapar Institute of Engineering and Technology, Patiala 147004, Punjab, India

E-mail: hbhunia@thapar.edu

Abstract. Porous carbon monoliths have been synthesized using furfuryl alcohol as a carbon precursor in the nanocasting technique. The highest surface area of 1225 m^2g^{-1} was found for the carbon monolith exhibited the micropore volume of 0.85 cm^3 g^{-1}. The various functional groups have been obtained on the surface of the adsorbent which increases its basicity as confirmed from the XPS analysis. The dynamic CO_2 uptake was 1.0 mmol g^{-1} at 30°C using temperature swing adsorption set up under 12.5% CO_2 flow. The effect of various adsorption temperatures along with CO_2 concentrations on the adsorption capacity was also investigated. The increase in the CO_2 uptake capacity by rising the temperature from 30°C onwards to 100°C confirms the involvement of exothermic and physisorption process. Also, the experiments for five adsorption-desorption cyclic runs were conducted for the regeneration study, indicating the good stability and reusabilty of the samples.

1. Introduction

The CO_2 greenhouse gas emissions are very harmful to the environment which plays a major role to enhance global warming. The burning of fossil fuels is the main cause of the increasing CO_2 emissions. The estimated CO_2 concentration in the year 2100 is 570 ppm [1] which will be the major threat to the human life. Therefore, the reduction in the CO_2 levels is very necessary to solve the environmental problems.

Porous carbons are considered as well known and effective candidates for the CO_2 capture. It exhibits the wide availability, low cost, and fast adsorption kinetics as compared to other materials. The different synthesis processes are implicated for the synthesis of porous carbons like direct carbonization, nanocasting, sol-gel, physical and chemical activation. Among the various methods, nanocasting is known as the best suitable technique due to the possibilities of varying the porosity/surface area of the carbons. It includes the precursor impregnation in the template, carbonization and then elimination of template [2].

The carbons in the form of the macroscopic level (monolithic form) have lot of advantages as compared to its powder form. The monoliths (meso and microporous) exhibits a lot of significant properties such as easy to deal, mass transfer, fast heat as well as high contacting efficiency [3-5] during the CO2 adsorption. The presence of micropores trapped the CO2 molecules effectively whereas, mesopores helps for the diffusion of CO2 in low resistant pathways[6]. In this regard, porous monolithic carbons (single-rock like) are developed and used for the CO2 adsorption studies under dynamic conditions. The pore size distribution curves and HRTEM analysis shows the presence of micropores. Whereas, the XPS analysis are confirmed the presence of the different functionalities on the adsorbent surface.

2. Experimental

2.1 Synthesis

The schematic diagram (Figure 1a) shows the preparation of meso/microporous carbon monoliths. For the synthesis of the monolithic silica template, 9.54 X 10^{-4} M of polyethylene glycol was dissolved in the water (9 ml) and then nitric acid (30%) was added slowly. The 8.1 ml of tetraethylorthosilicate and octadecyltrimethylammonium bromide (0.2M) was taken in the resulted solution and kept for stirring (15 min). The solution was then taken in the microplates and heated for 48 h at the temperature of 40-50°C for the gel formation. The gel form monoliths were then kept in the NH_4OH solution (1M) at 80 °C overnight and washed thrice with HNO_3 (0.1 M) and H_2O each. The silica monoliths were obtained after drying at 50 °C (96 h) and then kept for calcination (550 °C, 5 h). The synthesized monoliths were then impregnated with furfuryl alcohol solution with oxalic acid along with trimethyl benzene (FA/oxalic acid -250, ratio in moles). The monoliths were treated at 50 °C and then 100 °C (24h) in the oven for polymerization. The obtained monoliths were carbonized at 950 °C (4 h) with the 2 °C/min heating rate. Finally, the carbon monoliths were acquired by the complete removal of the template (silica) by treating with hydrofluoric acid (10%, by volume).

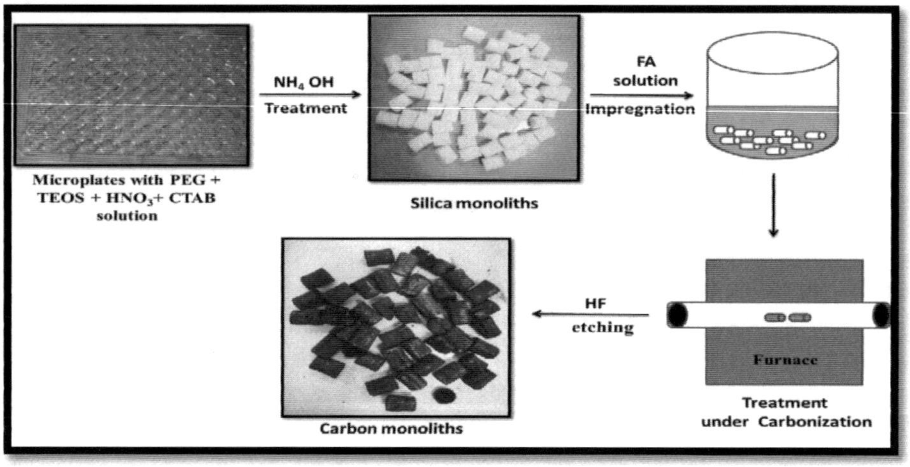

Figure 1. The schematic diagram for the preparation process of monolithic carbons

2.2 Apparatus

The Microtec Belsorp Mini–II, Japan instrument was used for the N_2 sorption isotherms and surface area for the samples. The morphology of the monoliths was examined by SEM (scanning electron microscope, JEOL JSM – 6510 LV) and HRTEM analysis (high-resolution transmission electron microscopy, JEOL JEM-2100). The investigation of various functional groups exists in the adsorbents was done by using X-ray photoelectron microscopy, PANALYTICAL X'Pert PRO. The CO_2 capture experiments were carried out using custom made temperature swing adsorption system equipped with the online gas chromatograph of Agilent 7890A. The CO_2 adsorption capacity, q_t (mmol g^{-1}) estimated by using Eq. (1). In which, m is the adsorbent weight (g), t (min) denotes the time in min, Q is the CO_2 flow rate in ml min^{-1} and C_0 and C are the inlet and outlet CO_2 concentrations.

$$q_t = \frac{1}{m}\int_0^t Q\,(C_o - C\,)\,dt \qquad (1)$$

3. Results and discussion

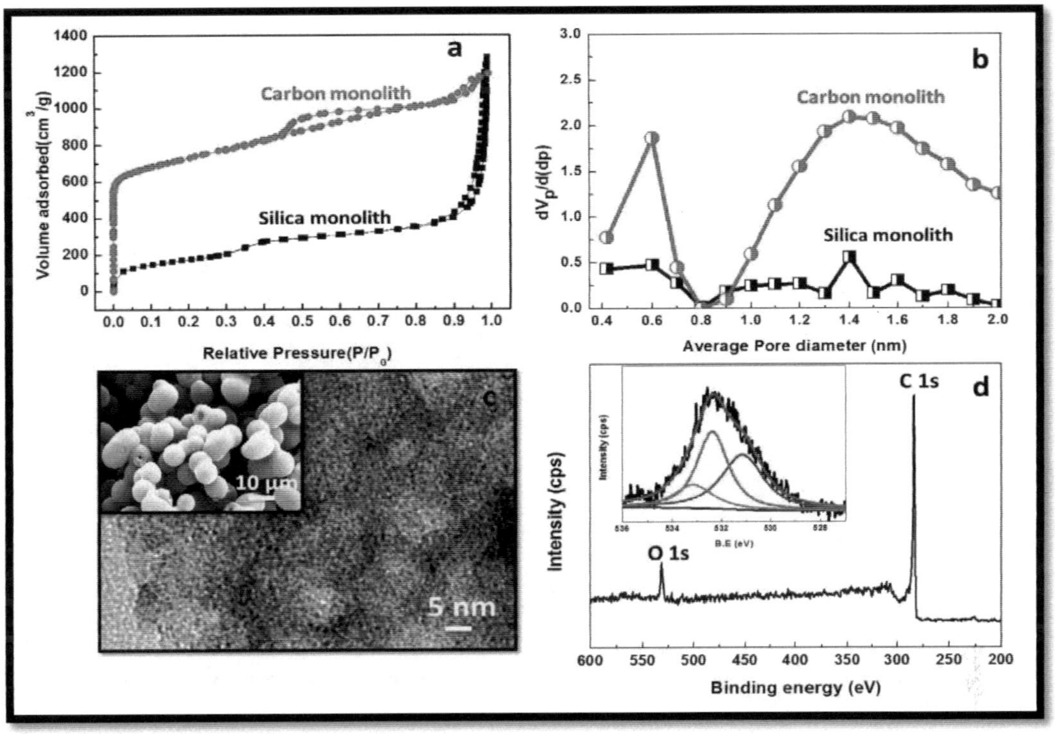

Figure 2. (a) N_2 sorption isotherms and (b) the pore size distribution curves for the monoliths; (c) HRTEM micrograph inset with the SEM image and (d) XPS spectra for the carbon monolith (inset: de-convoluted spectra of the O1s region).

Table 1. The adsorption capacity exhibited by the carbon monolith at different conditions

CO_2 concentration (%)	CO_2 adsorption (in mmol g^{-1})			
	30 °C	50 °C	75 °C	100 °C
5	0.33	0.21	0.18	0.15
7.5	0.52	0.34	0.26	0.22
10	0.79	0.55	0.35	0.28
12.5	1.0	0.72	0.57	0.49

The N_2 adsorption-desorption isotherms for the monolithic silica and carbon samples are shown in figure 2a. The silica monolith shows the type IV isotherm whereas, carbon monolith shows the type I isotherm which confirms the presence of micropores. The specific surface area for the silica and carbon monolith is 1225 and 760 m^2g^{-1}, respectively. The pore size distribution curves also indicate the existence of pores lies in between 1-2 nm (Figure 2b). The HRTEM image shows the microstructure of the spherical particles of the carbon monoliths which indicating its existence in the nanometer range (Figure 2c). Whereas, XPS spectra (Figure 2d) of the monolithic carbon confirming the existence of C1s and O1s region without the presence of any impurity. The de-convoluted peaks for the O1s region having the binding energies of 531.2eV, 532.3eV and 533.2eV assigned to C=O, C-OH groups, and oxygen in the water, respectively. Also, the C-OH groups increases the surface basicity [7] which can be beneficial to capture CO_2 molecules during adsorption.

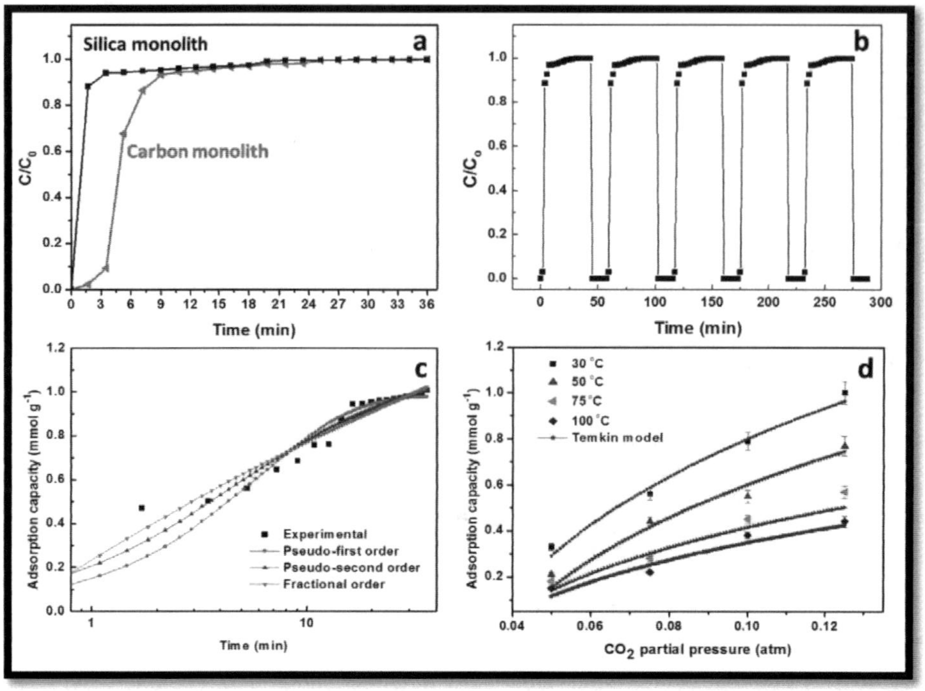

Figure 3. (a) CO_2 Breakthrough curves for the silica and carbon monolith; (b) CO_2 adsorption-desorption cycles for the carbon monolith; (c) CO_2 uptake kinetics; and (d) Fitting of the experimental data with the Temkin model at various adsorption temperatures

The breakthrough curves for the monoliths at 30 °C are shown in figure 3a. The maximum adsorption capacity (1.0 mmol g^{-1}) was exhibited by the carbon monolith whereas; silica monolith induced the capacity of 0.26 mmol g^{-1}. This is due to the higher textural properties (S_{BET}-1225 m^2g^{-1}) of the carbon monolith as well as the presence of basic oxygen functional groups exist on the surface. The CO_2 uptake exhibited by the carbon monolith at various temperatures as well as CO_2 concentrations is shown in table 1. It was seen that the adsorption capacity decreased by increasing the temperature indicating the involvement of physisorption as well as the exothermic process. However, the CO_2 uptake capacity increases by increasing the CO_2 concentration from 5 to 12.5% due to the enhancement of driving force by increasing the concentrations of CO_2. The experiment runs for the CO_2 adsorption-desorption cycles for carbon monolith at 30 °C under 12.5% CO_2 concentration can be seen in figure 3b. The quick desorption of CO_2 after the adsorption process can be estimated by the sudden fall of C/C_o after the complete adsorption. The adsorbents showed the excellent recyclability for CO_2. Therefore, it can be said that the synthesized monoliths can be effective as well as reused for multiple times for the CO_2 capture applications. The fractional order model fitted best with the experimental data and R^2 value is 0.99 (Figure 3c) using the Eqn.2. Figure 3d shows the fitting of experimental results with the Temkin isotherm model which exhibits the R^2 value of 0.99 using the Eqn.3. This indicates the heterogeneous surface of carbon monoliths.

$$q_t = q_e - \frac{1}{\left[\left((n-1)k_n/m\right)t^m + \left(1/q_e^{n-1}\right)\right]^{1/n-1}} \quad (2)$$

$$q_e = B\ln\left(K_T P\right) \quad (3)$$

q_t and q_e are the adsorption capacity (mmol g^{-1}) at a particular time t (minutes) and at equilibrium respectively. k_n, m and n are the rate constant of fractional order model. K_T is the Temkin constant in atm^{-1}, T is the temperature in Kelvin, B is the constant represents the sorption heat.

IOP Conf. Series: Materials Science and Engineering **625** (2019) 012014 doi:10.1088/1757-899X/625/1/012014

4. Conclusions
Herein, monolithic carbons were prepared using the nanocasting technique. The obtained carbon monoliths showed the S_{BET} of 1225 m^2 g^{-1} and exhibited the maximum CO_2 uptake (1.0 mmol g^{-1}) which is about 3 times greater than the silica monolith (0.26 mmol g^{-1}). Also, the synthesized monoliths exhibited good stability after five recycle runs of adsorption-desorption. The Temkin isotherm model fitting shows the heterogeneity on the surface of the monoliths. Thus, the synthesized monoliths in the present study can be used in the practical applications in CO_2 separation from flue gas in the coal fired power plants.

Acknowledgments
The authors are grateful to Thapar Institute of Engineering & Technology for providing the instrumental facilities.

First name	Surname	Institution	Email address	Title (Prof/ Dr./student)	Research Interests
Jasminder	Singh	*Thapar Institute of Engineering and Technology, Patiala 147004, Punjab, India.*	jasminder.singh91@gmail.com	Mr. (Ph.D. scholar)	Synthesis of porous materials
Soumen	Basu	*Thapar Institute of Engineering and Technology, Patiala 147004, Punjab, India.*	*soumen.basu@thapar.edu*	Dr. (Associate Professor)	Preparation of metal oxides porous carbons for CO_2 capture.
Haripada	Bhunia	*Thapar Institute of Engineering and Technology, Patiala 147004, Punjab, India.*	*hbhunia@thapar.edu*	Dr. (Professor)	CO_2 capture and utilization, Biodegradable polymer

References

[1] Tiwari D, Goel C, Bhunia H, Bajpai PK 2016 Novel nanostructured carbons derived from epoxyresin and their adsorption characteristics for CO_2 capture. *RSC Adv.* **6** 97728-38

[2] Tiwari D, Bhunia H, Bajpai PK 2017 Synthesis of nitrogen enriched porous carbons from ureaformaldehyde resin and their carbon dioxide adsorption capacity. *J CO_2 UTIL.* **21** 302-13

[3] El Kadib A, Chimenton R, Sachse A, Fajula F, Galarneau A, Coq B 2009 Functionalized inorganic monolithic microreactors for high productivity in fine chemicals catalytic synthesis. *ANGEW CHEM* **121** 5069-72

[4] Xin W, Song Y 2015 Mesoporous carbons: recent advances in synthesis and typical applications *Rsc Adv.* **5** 83239-85

[5] Hong WY, Perera SP, Burrows AD 2015 Manufacturing of metal-organic framework monoliths and their application in CO_2 adsorption. *Micropor. Mesopor. Mat.* **214** 149-55

[6] Wang J, Huang L, Yang R, Zhang Z, Wu J, Gao Y, et al 2014 Recent advances in solid sorbents for CO_2 capture and new development trends. *Energy Environ. Sci.* **7** 3478-518

[7] Tiwari D, Goel C, Bhunia H, Bajpai PK 2017 Melamine-formaldehyde derived porous carbons for adsorption of CO_2 capture. *J ENVIRON MANAGE* **197** 415-27

Electrochemical determination of diethylstilbestrol by using a magnetic nanoparticle/graphene composite film electrode

Yingyot Poo-arporn[1], Saithip Pakapongpan[2] and Rungtiva P. Poo-arporn[3,*]

[1] Synchrotron Light Research Institute, 111 University Avenue, Nakhon Ratchasima 30000, Thailand
[2] Thailand Organic and Printed Electronics Innovation Center, National Electronics and Computer Technology Center, NSTDA, Pathum Thani 12120, Thailand
[3] Biological Engineering Program, Faculty of Engineering, King Mongkut's University of Technology Thonburi, Bangkok 10140, Thailand

Email: rungtiva.pal@kmutt.ac.th

Abstract. The magnetic nanoparticle-graphene modified magnetic carbon electrode (MCE) was fabricated and the electrochemical effect of diethylstilbestrol (DES) on this electrode was studied by cyclic voltammetry (CV) and differential pulse voltammetry (DPV). The factors that affected the performance of the sensor were also optimized. The experimental results exhibited good electrocatalytic activity and a fast response for the electrochemical oxidation of DES. The linear range for the determination of diethylstilbestrol was 1 to 100 μM and the detection limits were 1.06 x 10^{-7} M based on S/N = 3.

1. Introduction

Diethylstilbestrol (DES) is a synthetic non-steroidal estrogen substance and its molecular structure contains two hydroxyl groups in the phenol-ring. Since the pharmacological and therapeutic effects of DES are the same as those of natural estrogen, it has been widely used in treatment of estrogen-deficiency disorders, and livestock production to promote growth [1]. However, in contrast to natural estrogen, DES is more stable and remains in the body longer [2-3]. Recently, it has been found that DES is one of the strongest potential carcinogens [4]. Therefore, the use of growth-promoting drugs for fattening livestock has been banned in many countries [5]. Based on this fact, there is an urgent need for a fast and efficient method for the determination of residues in animal tissues as well as DES in milk samples to prevent illegal usage.

Several methods have been used for determining DES [6-8]. Until now, most of the analytical methods towards this determination have required expensive instruments, trained personnel, and are very time-consuming. This added to the cost and complexity of the assay. Consequently, the method provides a convenient, simple, rapid, and economic means of detecting these food contaminants. DES has the property of electrochemical activity because of the hydroxyl groups in its molecules. Therefore, electrochemical techniques, which are often simple and less expensive, have been used to determine DES [9-13]. However, a major disadvantage of the method is its low sensitivity due to the slow electron transfer at unmodified electrodes. Interest has been focused on the development of chemically modified electrodes for DES detection. Interestingly, so far, a magnetic nanoparticle-graphene composite has not been developed for the electrocatalytic detection of these food contaminants. In this work, a new disposable electrochemical sensor for the fast determination of DES is fabricated based on the use of a magnetic carbon electrode (MCE) modified with magnetic nanoparticles and graphene composites.

2. Materials and methods

All electrochemical experiments were carried out with a Potentiostat (Metrohm Autolab PGSTAT302N, Ecochemie, Netherlands) in a three-electrochemical cell form using a magnetic carbon electrode (MCE) as a working electrode, a platinum wire as a counter electrode, and Ag/AgCl saturated KCl as the reference electrode.

The fabrication of Fe_3O_4-Gr composites was described elsewhere [14-15]. Briefly, 20 mg graphene oxide in 60 mL water was ultrasonicated for 3 h, and then 100 mg EDC and 80 mg NHS were added together in the solution of graphene. The mixture was stirred for 30 min and ultrasonicated for another 30 min to form a homogenous suspension. Next, 20 mg NH_2-Fe_3O_4 NPs were added into the suspension solution and the mixture was subjected to ultrasonication for 30 min. The reaction was carried out at 80 °C for 1 h under stirring. The Fe_3O_4-Gr composites were obtained by magnetic separation and washed with water several times. The obtained Fe3O4-Gr composites (1 mg/2.5 mL) were re-dispersed in DI water before further use. Finally, 5 µl of composite solution was dropped onto the MCE surface.

3. Results and discussion

3.1 Electrochemical characterization of Fe_3O_4-Gr composite-modified MCE

The electrochemical properties of the composite were evaluated through cyclic voltammetry (CV). The cyclic voltammograms of the Fe_3O_4-Gr composite-coated MCE in phosphate buffer solution at pH 7.0 at various scan rates can be seen in Figure 1. The anodic and cathodic peak currents are linear with potential scan rates ranging from 25 to 400 mV/s (inset in Figure 1). The results confirmed that composites have good electrochemical reversibility. Both the anodic peak current and the cathodic peak current were proportional to the scan rate, again demonstrating that the redox process of the Fe_3O_4-Gr composite on to the MCE was a surface-confined process.

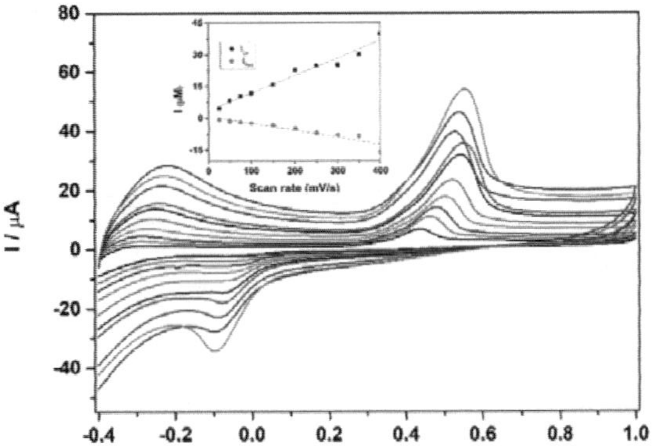

Figure 1. CVs of the Fe_3O_4-Gr/MCE in 0.1 M PBS at pH 7.0 containing 0.3 mM DES at different scan rates from inner to outer: 25, 50, 75, 100, 150, 200, 250, 300, 350 and 400 mV/s. Inset: The plot of the peak current vs. scan rates.

3.2 Optimization of DES monitoring

The accumulation time was determined in order to find the optimal operational condition for DES measuring. Thus, a plot of differential pulse voltammetric current versus accumulation time was made as shown in Figure 2. The current changed with a decrease in accumulation time from 5 to 210 seconds. Hence, the accumulation time of 5 seconds was selected as the optimized monitoring accumulation time, due to its good sensitivity.

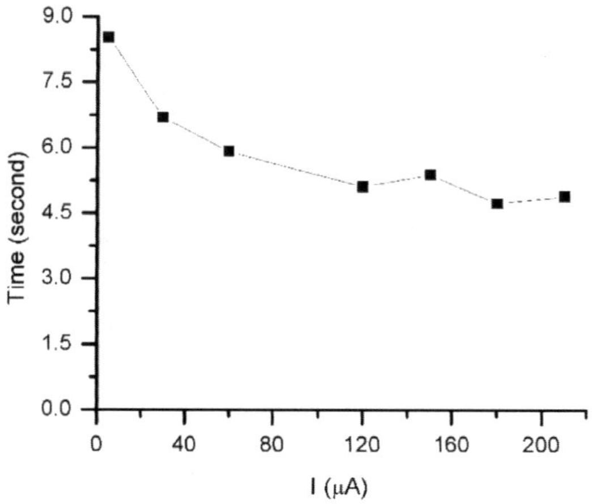

Figure 2. Effect of accumulation time on DPV peak current in 0.1 M PBS at pH 7.0 containing 0.3 mM DES.

The influence of pH on the sensor response is shown in Figure 3. With an increase of buffer pH, the CV response of the sensor shows no significant change until pH 8.0. A further increase of buffer pH led to a decrease in the response. Due to pH 7.0 being the normal pH in human and animal biological fluids and also the sensitivity of this sensor, pH 7.0 was chosen for use as the supporting electrolyte in all subsequent analytical experiments.

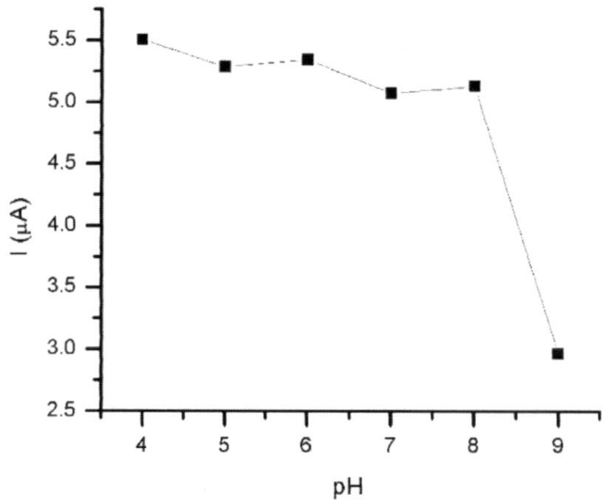

Figure 3. Effect of pH on CV peak current in 0.1 M PBS containing 0.3 mM DES at a scan rate of 50 mV/s.

3.3 Current response to DES

The current response of Fe_3O_4-Gr composite on to the MCE was investigated under the optimal experimental conditions as shown in Figure 4.

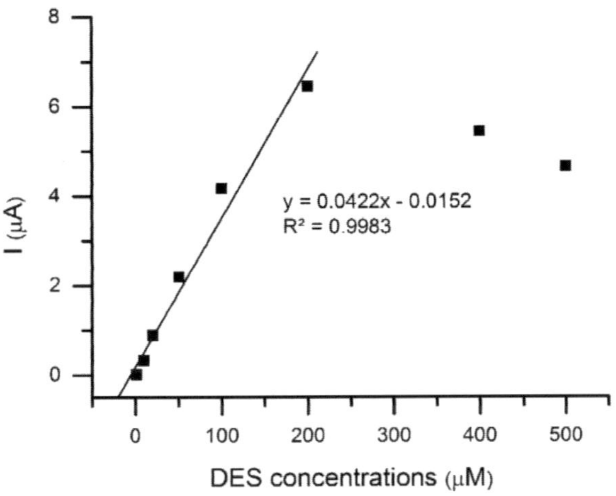

Figure 4. Calibration curve between the peak current and DES concentration from 1 to 500 μM in 0.1 M PBS at pH 7.0.

Upon addition of DES, the DPV response increased, indicating the fast response and high sensitivity of the electrode to DES. The calibration curve of the sensor is shown as an inset in Figure 4. The response to DES was linear in the range of 1 to 100 μM, and the sensitivity of this biosensor corresponding to the linear range was 42.2 μA/mM. The regression equation was Y = 0.0422x - 0.0152 with a correction coefficient of 0.9983, and the detection limit was 0.106 μM when the signal to noise ratio was 3.

4. Conclusions

This work demonstrates a new disposable sensor for DES based on Fe_3O_4-Gr composite modified on the magnetic carbon electrode. The electrochemical measurement showed good electrocatalytic activity and sensitivity, along with a wide range and low detection limit of the sensor to DES. This sensor fabrication could be employed for other applications of several analytes in food contaminant analysis.

Acknowledgments

This work was supported by the Asia Research Center, Chulalongkorn University; the research subsidy fund of fiscal year 2019 from King Mongkut's University of Technology Thonburi.

References

[1] Heitzman J. R. and Steroid J. 1979 *Biochem.* **11** 927
[2] Giusti M. R., Iwamoto K. and Hatch E. E. 1995 *Ann. Intern. Med.* **122** 778
[3] Tapiero H., Ba N. G. and Tew D. K. 2002 *Biomed. Pharmacother.* **56** 36
[4] Zhang L. Q., Li J, Ma T. T. and Zhang T. Z. 2008 *Food Chem.* **111** (2008) 498
[5] Xu Q., Wang M., Yu Q. S., Tao Q. and Tang M. 2011 *Analyst* **136** 5030

IOP Conf. Series: Materials Science and Engineering **625** (2019) 012015 doi:10.1088/1757-899X/625/1/012015

[6] Zhao D. C., Ji S. Y., Shao L. Y., Jiang M. X. and Zhang X. H. 2009 J. Chromatogr. A **1216** 7546

[7] Liu F. S., Wu P. X., Xie H. Z., Lin C. X., Guo Q. L., Yan C. and Chen N. G. 2005 Electrophoresis **26** 2342.

[8] Wang J., Ye Z. H., Jiang Z., Chen S. N. and Huang L. J. 2004 *Anal. Chim. Acta* **508** 171

[9] Ngundi M., Sadik O., Yamaguchi T. and Suye S. 2003 *Electrochem. Commun.* **5** 61

[10] Bin Q., Wei W., Chi W. Y. and Chen N. G. 2005 *Anal. Biochem.* **336** 196

[11] Lu D., Lin S., Wang L., Shi X., Wang C. and Zhang Y. 2012 *Electrochim. Acta* **85** 131

[12] Yu C., Ji W., Wang Y., Bao N. and Gu H. 2013 *Nanotechnology* **24** 115502

[13] Ma X. and Chen M. 2015 *Sens.Actuators B* **215** 445

[14] Pakapongpan S. and Poo-arporn P. R. 2017 *Mater Sci Eng C Mater Biol Appl.* **76** 398

[15] Pakapongpan S., Tuantranont A. and Poo-arporn P. R. 2017 *Sci. Rep.* **7** 12882

Oxalate precipitation synthesis of copper-manganese nanoparticles for methanol steam reforming

Y D G Edañol[1], R L Barbosa[2], S C Buenviaje[1], and L M Payawan Jr[3]

[1]Natural Sciences Research Institute, University of the Philippines, Diliman, Quezon City, 1101, Philippines
[2]Department of Chemistry and Biochemistry, Lamar University, 4400 MLK Blvd, Beaumont, TX 77710, USA
[3]Institute of Chemistry, College of Science, University of the Philippines, Diliman, Quezon City, 1101, Philippines

Email: yasmin_501_96@yahoo.com.ph

Abstract. CuMn spinel nanoparticles were prepared using the oxalate precipitation synthesis technique using nitrate salts as precursors. The samples were characterized using X-ray diffractometry (XRD), and field emission – scanning electron microscopy (FE-SEM). The mixed oxalate precursor, CuMn*C2O4, was found to be single-phase and crystalline compound. H2-assisted temperature-programmed decomposition of the mixed oxalates yielded Cu1.5Mn1.5O4 spinel nanoparticles with CuO present in very minute amounts. FE-SEM analysis of the Cu1.5Mn1.5O4 spinel showed a spherical morphology of the nanoparticles. The average size of the Cu1.5Mn1.5O4 spinel nanoparticles was 11.512 nm. Thus, the nano-sized Cu1.5Mn1.5O4 spinel nanoparticles can be used as catalysts for the methanol steam reforming (MSR) reaction.

1. Introduction

In times where environmental problems pose a great threat to the society, development of cleaner and new sources of energy is of utmost concern [1]. The Fischer-Tropsch (FT) reaction was developed by Franz Fischer and Hans Tropsch in the 1920s [2]. The reaction involves the use of synthesis gas, a gaseous mixture of carbon monoxide (CO) and hydrogen (H_2), to produce sulfur-free organic products at operating temperatures of 200-350°C. The synthesis gas (CO/H_2), which can be harnessed from natural gas, coal, biomass or any carbonaceous source, makes the synthesis of a wide range of organic compounds such as paraffins (1), olefins (2), and oxygenates (3) is possible [3-4]. The most important reactions are summarized below:

$$\mathbf{(2n + 1)\ H_2 + nCO \rightarrow C_nH_{2n+2} + nH_2O \quad (1)}$$
$$\mathbf{(2n)\ H_2 + nCO \rightarrow C_nH_{2n} + nH_2O \quad (2)}$$
$$\mathbf{(2n)\ H_2 + nCO \rightarrow C_nH_{2n+1}OH + (n\text{-}1)\ H_2O \quad (3)}$$

Hydrocarbons and alcohols can be used for reforming reactions for the in situ production of hydrogen fuel [5-6]. Although industrial production of hydrogen mainly utilizes methane as fuel, other fuels such as methanol presents several advantages compared to methane and all other fuels. Aside from being the simplest alcohol, reforming with methanol occurs at low temperatures (200-300°C) due to the absence of a strong C-C bond. Methane is reformed above 500°C [7] while ethanol, another common fuel in hydrogen production, is reformed at around 400°C [6]. Other advantages of using methanol is its high miscibility in water, biodegradability, liquid state at atmospheric conditions, and high hydrogen to carbon ratio. The only problem with methanol is its toxicity. The methanol steam reforming (MSR) reaction is shown in equation 4:

$$\mathbf{CH_3OH + H_2O \rightleftharpoons CO_2 + 3H_2 \quad \Delta H = 49.7\ kJ/mol \quad (4)}$$

The MSR reaction mainly produces hydrogen but other products that are formed must be taken into consideration. There are two side reactions that occur simultaneously with the MSR reaction:

$$\mathbf{CH_3OH \rightleftharpoons CO + 3H_2 \quad \Delta H = 90.2\ kJ/mol \quad (5)}$$
$$\mathbf{CO + H_2O \rightleftharpoons CO_2 + H_2 \quad \Delta H = -41.2\ kJ/mol \quad (6)}$$

Altogether, the MSR, the methanol decomposition reaction (5) and water-gas shift (WGS) reaction (6) yields a reaction mixture composed of unreacted methanol, water, hydrogen, carbon dioxide, and minute amounts of carbon monoxide. If the hydrogen produced will be used as fuel in polymer electrolyte membrane fuel cells (PEMFC), carbon monoxide formation must be suppressed [1]. This can be achieved by tuning the performance of the catalyst. The catalyst must be highly active and highly selective to produce large amounts of hydrogen and negligible amounts of carbon monoxide. Among the catalysts for the MSR reaction, copper-based catalysts possess both high activity and high selectivity [8-9]. However, these catalysts are pyrophoric and are deactivated by thermal sintering. Addition of promoter metals further improve the activity of the Cu-based catalyst. Furthermore, method of preparation affects the overall properties of the catalyst [10].

Oxalate co-precipitation synthesis is an effective catalyst preparation technique for MSR catalysts. Copper-based catalysts that were produced via the oxalate precipitation technique have high dispersion, smaller particle and crystallite size, and high surface areas [1]. In this study, copper-manganese (CuMn) nanoparticles for the MSR reaction were synthesized via the oxalate precipitation technique. The CuMn nanoparticles were characterized using), x-ray diffractometry (XRD), and field-emission scanning electron microscopy (SEM).

2. Methodology

2.1 Oxalate precipitation synthesis of copper-manganese mixed oxalates

Copper (II) nitrate hexahydrate and manganese (II) nitrate hexahydrate, with a 1:3 ratio were mixed in acetone. Accompanied with vigorous stirring, an excess of oxalic acid was added to complete the co-precipitation. After 3 hours of stirring, the wet oxalate cake was centrifuged and the excess acetone was discarded. The solid products, labeled as CuMn*C_2O_4, were grounded and then dried at 110°C for 24 hours.

2.2 H2-assisted temperature-programmed decomposition of the mixed oxalates

The temperature-programmed decomposition (TPDec) of the mixed CuC_2O_4 and MnC_2O_4 was performed from room temperature up to 500°C with a ramp rate of 6°C/min under a flow of 10% H_2 in argon. The total flow rate of the gas mixture was at 30 mL/min. After cooling to room temperature, the products were collected and labeled as CuMn.

2.3 X-ray diffractometry

Both CuMn*C_2O_4 and CuMn were characterized using a Rigaku X-ray Diffractometer (Cu Kα = 1.542Å) at the Lamar University. The X-ray tube was set at a voltage of 40.0 kV and a current of 30.0 mA with 2θ scan range 10° to 90°.

2.4 Field emission - scanning electron microscopy (FE-SEM)

CuMn*C_2O_4 and CuMn samples were mounted in carbon tape and were aspirated using compressed air. Morphology of the samples was studied using a Hitachi SU8230 field emission-scanning electron microscope (FE-SEM) housed at the Department of Mining, Metallurgical, and Materials Engineering

(DMMME), University of the Philippines, Diliman. The accelerating voltage was set to 15.0 kV with probe current at 21 pA.

3. Results and discussion

3.1 Oxalate precipitation synthesis of copper-manganese mixed oxalates

Bimetallic CuMn catalysts were prepared via the oxalate precipitation route. In this technique, double-chelation of the oxalate ligands with both metals form a metal-organic framework type of structure [11]. This is indicated in Scheme 1.

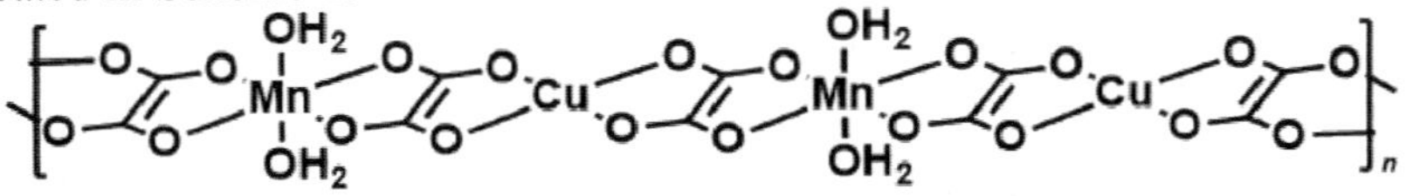

Scheme 1. Copper-Manganese Mixed Oxalate Structure

The mixed oxalate products are catalytically inactive. To produce catalytically active catalysts, the oxalate ligand must be thermally stripped off [11]. The H_2-assisted temperature programmed-decomposition of the mixed oxalates can proceed via the two equations below:

$$\mathbf{MeC_2O_4 \rightarrow Me + 2CO_2} \quad \textbf{(Eqn. 7)}$$
$$\mathbf{MeC_2O_4 \rightarrow MeO + CO + CO_2} \quad \textbf{(Eqn. 8)}$$

Equations 7 and 8 shows us that either the metal or the metal oxide is formed depending on the gaseous products generated. The decomposition of the pure copper oxalate occurs at 265°C [11] while the pure manganese oxalate decomposes at 303°C [12]. The decomposition of the mixed oxalate occurs at around 280°C. The chemical environment during the decomposition and the nature of the metal determines whether the metal oxalate is reduced to a pure metallic phase. The thermal decomposition formed under 10% H_2 diluted in Ar, a reducing atmosphere, was expected to produce a pure metallic Cu phase. Meanwhile, complete reduction Mn-oxalates or oxides to pure metallic states is impossible. Thus, the expected product is a pure metallic Cu phase dispersed in a matrix of Mn oxides [13].

3.2 X-ray diffractometry

X-ray diffractometry was used to identify the identity, as well as the structure of the synthesized oxalate precursors and the thermally decomposed products. Figure 1a shows the obtained XRD pattern of copper oxalate, manganese oxalate, CuMn*C_2O_4, and CuMn. The CuMn*C_2O_4 XRD pattern possesses peaks from the individual oxalates, CuC_2O_4 and MnC_2O_4. Square symbols indicate peaks from CuC_2O_4 while circular symbols indicate peaks from MnC_2O_4. The appearance of these peaks suggest that the product was a single-phase mixed oxalate. The formation of the single-phase mixed oxalate might be due to the isomorphous substitution between the manganese and copper oxalate phases. The single-phase product is expected to form a well-defined CuMn spinel oxide [13].

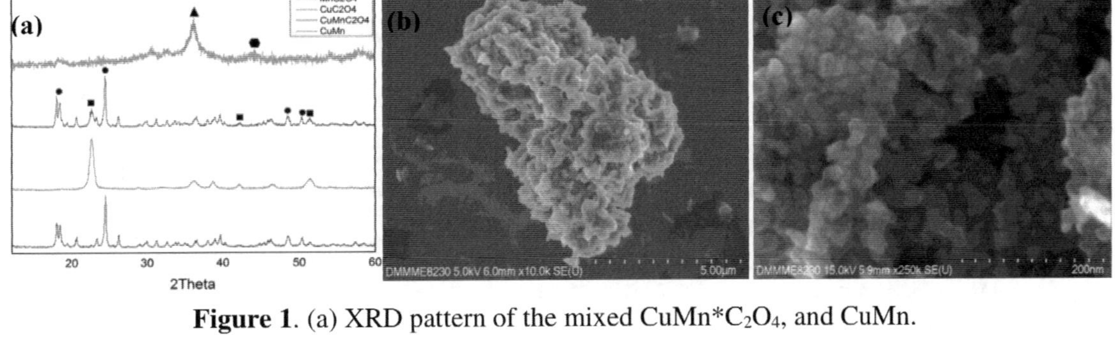

Figure 1. (a) XRD pattern of the mixed CuMn*C_2O_4, and CuMn.

XRD profile of the thermally decomposed product, CuMn, suggest that the CuMn exist as a $Cu_{1.5}Mn_{1.5}O_4$ spinel phase together with a minor CuO phase [13]. The triangular symbol denote the peak specific for the $Cu_{1.5}Mn_{1.5}O_4$ spinel while the hexagonal symbol denotes the peak for the CuO phase. The $Cu_{1.5}Mn_{1.5}O_4$ spinel is a non-stoichiometric form of $CuMn_2O_4$ and is commonly denoted by $Cu_xMn_{3-x}O_4$ or $Cu_{1+x}Mn_{2-x}O_4$. It is important to note the absence of the diffraction peaks for other crystalline Mn oxides species, such as Mn_2O_3 or Mn_3O_4, from the XRD profile of CuMn. This suggests that Mn_2O_3 or Mn_3O_4 are highly dispersed in the CuMn or the oxides are amorphous in nature [14].

3.3 Field emission – scanning electron microscopy (FE-SEM)

Morphological and particle size studies of the mixed $CuMn*C_2O_4$ and CuMn are essential because the catalytic activity and performance are highly influenced by particle size. Figure 1b and 1c shows the FE-SEM images of the mixed $CuMn*C_2O_4$ and CuMn.

It can be observed from Figures 1b and 1c that the mixed oxalates are crystalline in nature but the morphology of the crystal is hard to assess. The apparent crystalline structure agrees with the XRD results. Figure 1c shows the spherical morphology of CuMn which is actually a $Cu_{1.5}Mn_{1.5}O_4$ spinel. Particle size analysis using ImageJ software revealed that the particles have an average particle size of 11.512 nm. A similar CuMn spinel has been synthesized using the water-free soft reactive grinding synthesis technique and the average size was found to be at 17 nm [13]. Thus, the synthesized $Cu_{1.5}Mn_{1.5}O_4$ spinel particles can be used as catalysts for the MSR reaction. Furthermore, the smaller sizes of the particles suggest higher active surface area and higher performance in the MSR reaction.

4. Summary

Oxalate precipitation technique was employed for the preparation of the catalytically inactive mixed oxalate $CuMn*C_2O_4$, and the catalytically active $Cu_{1.5}Mn_{1.5}O_4$ spinel nanoparticles. XRD profile of $CuMn*C_2O_4$ consists of peaks from the individual oxalates. Thus, $CuMn*C_2O_4$ is a crystalline single-phase mixed oxalate. FE-SEM supported the crystalline nature of the mixed oxalates. However, the morphology of the mixed oxalate was difficult to assess. XRD profile of CuMn reveals that the thermally decomposed product consists of $Cu_{1.5}Mn_{1.5}O_4$ spinel and a minor phase of CuO. This confirms that Mn oxides cannot be completely reduced to metallic Mn. The absence of Mn oxide peaks in the XRD profile suggest that the Mn oxides are highly dispersed or amorphous in nature. FE-SEM analysis of CuMn reveal the spherical morphology of the nanoparticles. The average size of the CuMn is 11.512 nm. The nano-range size of the $Cu_{1.5}Mn_{1.5}O_4$ spinel particles make them suitable catalysts for the MSR reaction.

Acknowledgements

The authors would like to thank the UP Natural Sciences Research Institute (Project Code: CHE-19-1-05) for the financial support. The authors would also like to thank the Laboratory of Inorganic Nanomaterials, Institute of Chemistry, University of the Philippines, Diliman for moral support.

References

[1] Sá, Sandra, et al. "Catalysts for methanol steam reforming—a review." Applied Catalysis B: Environmental 99.1-2 (2010): 43-57.

[2] Steynberg, A. P. "Introduction to fischer-tropsch technology." Studies in surface science and catalysis. Vol. 152. Elsevier, 2004. 1-63.

[3] Thomas, John Meurig, et al. "Principles and practice of heterogeneous catalysis." Angewandte Chemie-English Edition 36.23 (1997): 2689-2689.

[4] Gnanamani, Muthu Kumaran, et al. "Fischer–Tropsch synthesis: Activity of metallic phases of cobalt supported on silica." Catalysis today 215 (2013): 13-17.

[5] Damle, Ashok S. "Hydrogen production by reforming of liquid hydrocarbons in a membrane reactor for portable power generation—Model simulations." Journal of Power Sources180.1 (2008): 516-529.

[6] Tosti, S., et al. "Low temperature ethanol steam reforming in a Pd-Ag membrane reactor: Part 1: Ru-based catalyst." Journal of Membrane Science 308.1-2 (2008): 250-257.

[7] Chen, Yazhong, et al. "Hydrogen production capacity of membrane reformer for methane steam reforming near practical working conditions." Journal of Membrane Science322.2 (2008): 453-459.

[8] Papavasiliou, Joan, George Avgouropoulos, and Theophilos Ioannides. "Steady-state isotopic transient kinetic analysis of steam reforming of methanol over Cu-based catalysts." Applied Catalysis B: Environmental 88.3-4 (2009): 490-496.

[9] Huang, Chih-Yung, et al. "Performance of catalysts CuO-ZnO-Al2O3, CuO-ZnO-Al2O3-Pt-Rh, and Pt-Rh in a small reformer for hydrogen generation." Journal of Power Sources 166.2 (2007): 450-457.

[10] Karim, Ayman M., Travis Conant, and Abhaya K. Datye. "Controlling ZnO morphology for improved methanol steam reforming reactivity." Physical Chemistry Chemical Physics10.36 (2008): 5584-5590.

[11] Xiang, Yizhi, and Norbert Kruse. "Cobalt–copper based catalysts for higher terminal alcohols synthesis via Fischer–Tropsch reaction." Journal of energy chemistry 25.6 (2016): 895-906.

[12] Donkova, B., and D. Mehandjiev. "Mechanism of decomposition of manganese (II) oxalate dihydrate and manganese (II) oxalate trihydrate." Thermochimica acta421.1-2 (2004): 141-149.

[13] Liu, Qian, et al. "Waste-free soft reactive grinding synthesis of high-surface-area copper–manganese spinel oxide catalysts highly effective for methanol steam reforming." Catalysis Letters 121.1-2 (2008): 144-150.

[14] Morales, María Roxana, Bibiana P. Barbero, and Luis E. Cadús. "Total oxidation of ethanol and propane over Mn-Cu mixed oxide catalysts." Applied Catalysis B: Environmental67.3-4 (2006): 229-236.

IOP Conf. Series: Materials Science and Engineering **625** (2019) 012017 doi:10.1088/1757-899X/625/1/012017

Fabrication of solar cell using titanium, zinc, carbon and gold nanoparticles

A D C Z S Lim and M R M Julaihi

Swinburne University of Technology Sarawak, Jalan Simpang Tiga, 93350 Kuching Sarawak, Malaysia

Email: 100073170@students.swinburne.edu.my

Abstract. Gold, titanium dioxide, zinc oxide, and carbon nanoparticles each have characteristics that can be used as the active, electron transport, and hole transport layer in a solar cell. Nanoparticle based solar cells have emerged as a good substitute for silicon solar cells. They cost lower to manufacture, as compared to producing solar cells with crystalline semiconductor material. This is because the nanoparticles can be synthesised at room temperature, while silicon solar cells need to be made using high temperature vacuum deposition method. In this research, the nanoparticles will be synthesised using submerged glow discharge plasma. After synthesis, the nanoparticle products were analysed using scanning electron microscopy to determine their size and shape. The solar cell will then be fabricated in multiple configurations. Testing on these multiple solar cells will be done under direct sunlight to determine if the cells work and how each material affects the cell.

1. Introduction

Silicon solar cells are commonly used as a source of renewable energy. They are reliable and have the highest efficiency among all other alternatives. Even so, nanoparticles-based solar cells have become a promising substitute. They have a lower manufacturing cost as compared to conventional silicon solar cell. This is because the nanoparticles can be synthesised at room temperature.

Recent studies have shown that metal nanoparticles such as gold nanoparticles absorb and scatter light based on localized surface plasmon resonance (LSPR) [1]. By incorporating these characteristics with other materials such as titanium dioxide, zinc oxide, and carbon, it is possible that a solar cell with an efficiency of 50% or more will emerge. This research focuses on using submerged glow discharge plasma (SGDP). This method is relatively new and has not been widely used in the synthesis of nanoparticles for solar cells fabrication.

2. Literature review

2.1 Submerged glow-discharge plasma setup (SGDP)

SGDP is a useful and straightforward because non-equilibrium plasma provides fast reactions at atmospheric pressure [2]. When an external energy source is connected to the system and power is supplied, the neutral gasses in the solution are ionized. Electrons are freed from their covalent bond due to the excess energy and plasma is generated. The plasma creates a strong electric field as it has an equal number of negative electrons and positive ions. The material that is to be reduced to nanoparticles will be connected to cathode, while the material that will supply electrons will be connected to the anode.

IOP Conf. Series: Materials Science and Engineering **625** (2019) 012017 doi:10.1088/1757-899X/625/1/012017

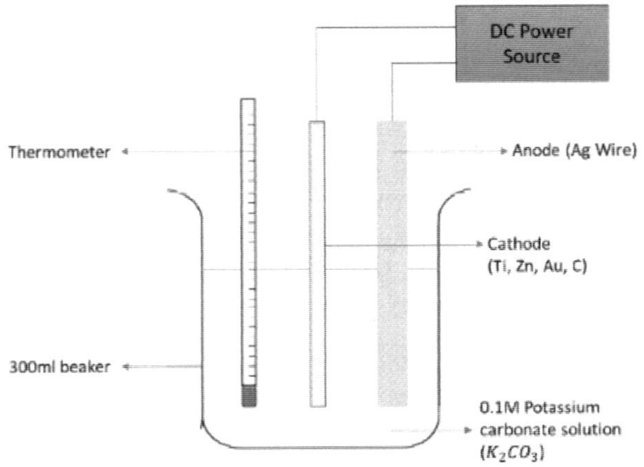

Figure 1. Submerged glow discharge plasma (SGDP) method

There are 4 stages of reaction using the SGDP; electrolysis, vaporization, transition, and full plasma stage [2]. During the full plasma stage, the high heat of the plasma partially melts the cathode surface to produce nanoparticles. After some time, as the current flow concentrates on the current spot, the plasma near here will be heated extra by added Joule heating and melts away or ionizes the rest of the cathode and the reaction continues as long as a stable voltage is supplied. Molten metal will cool down immediately in the electrolyte and form fine spherical particles.

Since the nanoparticles synthesized from SGDP have nearly uniform shape and size, it is highly advantageous to explore the possibility of fabricating an all-metal nanoparticle solar cell. The following sections describe the candidate nanoparticles and their role as solar cell materials.

2.2 Titanium dioxide nanoparticles

Titanium dioxide (TiO_2) is a dielectric material with its bandgap in the range of 3.0-3.2eV. This means it is a wide bandgap semiconductor that has strong absorption of high energy photons. It is also transparent to photons with energy (hv) that have bandgap energy (Eg) in the range of 1-2eV [3]. TiO_2 also has a high reflective index, up to n=2.5, that enhances the solar cell short circuit current density, J_{SC} [4]. Besides that, it can also be used as an electron transport layer (ETL) in the solar cell. $TiO_2$is currently widely used as ETL in perovskite solar cells (PSCs) due to its high electron mobility. This was observed in the case of P3HT:$PC_{61}BM$ perovskite was used as an active layer. The device that had no ETL showed optimum efficiency at just 1.94%, while the device that had TiO_2 ETL performed at 4.29% [4].

2.3 Zinc oxide nanoparticles

Zinc oxide (ZnO) has very good tuneable optoelectronic properties and is a widely used transparent conductive oxide. ZnO is used as a hole blocking layer for perovskite solar cells. It efficiently reduces the recombination of electron and hole pair at the perovskite/fluorine doped tin oxide (FTO) interface [5]. By reducing the recombination rate, more electrons will move towards the P-type side of the solar cell and thus more electricity is produced. The efficiency will then increase, such as in the case of a perovskite solar cell with glass/FTO/ZnO layers had a maximum efficiency of 14.2% [5]. ZnO, similar to TiO_2 is also widely studied and used as an electron transport layer. Both have similar bandgap value at 3.2eV, but ZnO has much higher electron mobility, around 115-155cm^2/Vs^2 [6].

2.4 Gold nanoparticles (Au NP)

Au NP has two great properties that helps increase the efficiency of a solar cell. They improve the absorption efficiency of the photovoltaic layer. Au NP that are placed above the photovoltaic layer that provides an angular spread which increases the effective light path and therefore, the absorption, this term is called light scattering. They can also facilitate light concentration. Au NP is embedded in the photovoltaic layer. The presence of an intense local field on the surface of the Au NPs increases the

IOP Conf. Series: Materials Science and Engineering **625** (2019) 012017 doi:10.1088/1757-899X/625/1/012017

sunlight absorption. Besides that, Au NP has a spectroscopic feature known as surface plasmon resonance, giving it good energy absorption in the visible light range. This feature comes from a collective resonant oscillation of the free electrons in its conduction band [7]. The absorption is also reliant on the shape and size of the particle. Spherical shaped particles 60nm and below showed much better light absorption than scattering. Besides that, a TiO_2-Au nanocomposite can be fabricated as an active layer due to their enhanced plasmonic resonance and light absorption.

2.5 Carbon nanoparticles (CNPs)

Solar cells fully made out of carbon nanoparticles have also been fabricated, but their efficiencies are somewhat subpar at 0.46% being the champion efficiency [8]. For this research, the CNPs will be used as the hole transport layer (HTL). Single walled nanotubes (SWNTs) have emerged as an effectively substitute to the commonly used PEDOT:PSS as the HTL between photoactive layer and indium tin oxide glass for perovskite solar cells. The results showed that 12.2nm thick layer of SWNTs had better performance as compared to PEDOT:PSS. Graphene oxide has also been predominantly used as OPV devices' HTL, but they are difficult to produce and are not cost effective. CNPs have been tested as an alternative to the expensive hole-transporter spiro-OMeTAD in the configuration of FTO/TiO_2 mesoporous $CH_3NH_3PbI_3$-xClx/CDs/Au [9].

A solar cell made of only metal nanoparticles have shown to work before, in the example of heterojunction solar cells that is made of di-copper oxide, zinc oxide and titanium oxide, and gold nanoparticles. They have shown to provide an efficiency of 0.25%, FF of 0.33, JSC of 2.7mAcm-2and VOC of 0.28V [10].

3. Nanoparticle synthesis

Figure 1 shows a 300ml glass beaker with 200ml of 0.1M potassium carbonate (K_2CO_3). Titanium/gold/zinc wire of 0.5mm in diameter and exposed length of 50mm, silver (Ag) wire/carbon rod of 1mm with a diameter of 1mm and same exposed length, and a thermometer are dipped in it. Titanium/gold/zinc/carbon wire is connected to the negative output of the DC power source, making it the cathode which will be dissolved. The Ag wire on the other hand is connected to the positive output, and thus making it the anode. The anode and cathode were cleaned by dipping them in 10% sulphuric acid for 10 seconds, then washed with distilled water.

The voltage is initially set at 5 volts to heat the electrolyte for 60 minutes. After that, the voltage is increased by steps of 5V every minute until the glow can be seen. The voltage is then maintained for the production of the nanoparticles. The voltage to reach glow discharge for this experiments materials are 95V for Zn, 120V for Ti, 100V for C, and 125V for Au.

3.1 Nanoparticle collection and analysis

After SGDP, the products will be collected by centrifuging for 10 minutes. The product will then be washed down with distilled water to remove the electrolyte. The nanoparticles will be sent for Scanning Electron Microscopy (SEM) to identify the elemental composition and map each type of nanoparticle.

4. Fabrication of solar cell

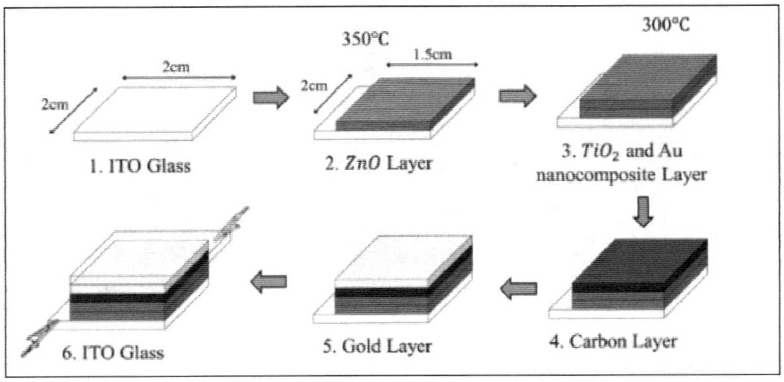

Figure 2. Proposed fabrication route for nanoparticles-based solar cell.

4.1 Step 1: Base preparation (Cathode)

The conducting side of ITO (indium tin oxide) glass can be determined by using a multimeter. The coated side is the one that shows low resistance. The coated side is faced upwards and a quarter of its surface area will be covered with adhesive tape.

4.2 Step 2: Electron transport layer (ETL) preparation

The electron transport layer (ETL) ZnO nanoparticles is deposited on the exposed surface and spread evenly using a dropper and a spin coater. The glass plate is placed on a hot plate at 350°C for 20 minutes. The hot plate is turned off while waiting 10 minutes before removal. Heat treatment is done to ZnO to increase the surface roughness and also particle surface area. The absorption of light in the 350-700nm wavelength region will be increased [11]. The ZnO layers that are sintered at 300-400°C showed a homogeneous distribution of closely packed, grouped nanoparticles. The heating has no effect on the size of the particle or the porosity of the layers. Oxide particles are more likely to grow at these temperatures and thus reducing the grain boundary.

4.3 Step 3: Active layer

The Ti-Au nanocomposite layer will be the active layer. Two to three drops of it will be added atop of the prepared ETL. The spin coater is used to evenly coat the layer. This layer should be around 30nm thick for maximum efficiency. The hot plate is warmed to 300°C and the glass plate sintered on it for 5 minutes. This is to remove any excess methanol or water that is present in the composite material.

4.4 Step 4: Hole transport layer (HTL) preparation

Carbon NPs are placed on the active layer to facilitate hole transport. The layer is made using the droplet and spinner but no additional heating is required. This layer will be harder to smoothen out as the NPs are coarse, thus the layer will be verified for smoothness after spin coating.

4.5 Step 5: Electrode layer

The Au layer is deposited on the HTL or the active layer and acts as the electrode for the cell, which helps to hold the holes that will recombine with the incoming electrons from the cathode. This layer is prepared similar to the previous steps. However, heating is not required.

4.6 Step 6: Top preparation (Anode)

After all layers have set and dried, another ITO glass plate with the conductive side facing down is then placed upon them. The ITO plates will be placed laterally as shown in Figure 2 to facilitate good electrical contact, and avoid short circuiting. Two binder clips are used to keep the ITO plates in place and the tape from step 1 is removed to expose the rest of the ITO. Even though Au is a very good conductor, this ITO plate is added so that no sunlight exposed area of the cell is blocked and power conversion is maximized, also protecting the Au layer from smudging during testing.

5. Results and discussion

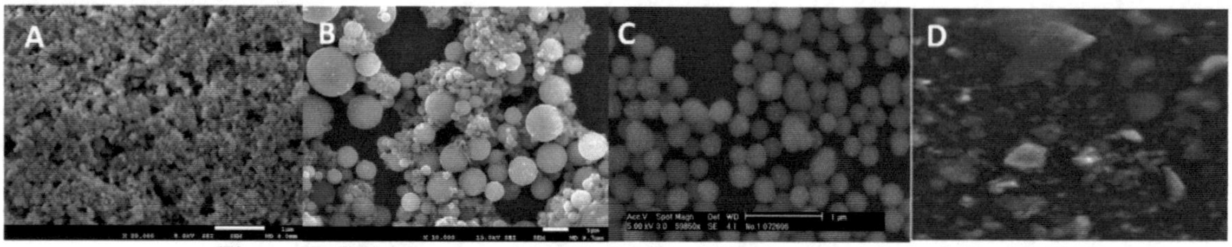

Figure 3. SEM images of (A)ZnO NPs (B)TiO_2 NPs (C) Au NPs (D)C NPs

The synthesized nanoparticles are shown in Figure 3. All nanoparticles were spherical shaped. Because of this shape, they will have good absorbance of light and improve the efficiency of the solar cell.

A prototype solar cell was fabricated using copper oxide, titanium oxide and zinc oxide nanoparticles sandwiched between two ITO glass plates measuring 20mm by 20mm. The prototype was exposed to sunlight at midday. To eliminate heating effect, the sunlight exposure was repeated several times in quick succession. Using a multimeter, the voltage measured was 0.25V, while the current was 0.005mA. When the prototype was removed from sunlight, the voltage and current measurements dropped to zero almost immediately. Although the photocurrent generated was small, the experiment has shown that the prototype solar cell could generate electrical current when exposed to sunlight. Further experimentation will be done on fabricating the solar cell to determine its feasibility and to determine the efficiency.

6. Conclusion

The proposed solar cell using a combination of titanium, zinc, carbon and gold nanoparticles have the potential to generate electrical current under sunlight exposure. The addition of carbon as the hole transport layer, and TiO_2-Au nanocomposite layer to exploit its plasmonic resonance would enhance the efficiency of the solar cell. Submerged glow-discharge method was used to produce the nanoparticles, and is found out to be a good substitute instead of using other synthesizing methods. Initial testing using a prototype solar cell has shown that the fabricated solar cell generated electricity mainly by photoelectric effect and not due to thermoelectric effect from the sun. The efficiency of the cell could be improved by using more sophisticated methods, or more refined nanoparticles, but overall this is a good stepping stone for future works. This type of solar cell is promising for future development not only because of it being economical but also due to its environmentally friendly synthesis method.

References

[1] Shi, X, Ueno, K, Oshikiri, T, Sun, Q, Sasaki, K and Misawa, H 2018 *Nat. Nanotechnol.*, **13** 953
[2] Yek, PNY, Julaihi, MRM, Osman, MS, Tiong, TC, Lee, WH and Lee, CL 2018 Submerged Glow-Discharge Plasma: An Economical Approach to Convert Construction Scrap Metal into Nanomaterials *E3S Web of Conferences* **34** 1
[3] Breda, C, Dehouchea, Z, Tahhana, A, Anson, T and Fern, GR *IEEE Phot. Spec. Conf.* doi: 10.1109/PVSC.2014.6925155
[4] Sun, L, Shen, W, Chen, W, Bao, X, Wang, N, Dou, X, Han, L and Wen, S 2014 *Thin Solid Films* **573** 134
[5] Hadouchi, W, Rousset, J, Tondelier, D, Geffroy, B and Bonnassieux, Y 2016 Zinc oxide as a hole blocking layer for perovskite solar cells deposited in atmospheric conditions *RSC Adv.* **6** 67715
[6] Kytin, V, Dittrich, T, Koch, F and Lebedev, E 2001 *Appl. Phys. Lett.* **79** 108
[7] Link, S and El-Sayed MA 2003 *Annu. Rev. Phys. Chem.* **54** 331
[8] Ramuz, MP, Vosgueritchian, M, Wei, P, Wang, C, Gao, Y, Wu, Y, Chen, Y and Bao, Z 2012 Evaluation of solution-processable carbon-based electrodes for all-carbon solar cells ACS Nano 6 10384

IOP Conf. Series: Materials Science and Engineering **625** (2019) 012017 doi:10.1088/1757-899X/625/1/012017

[9] Paulo, S, Stoica, G, Cambarau, W, Martinez-Ferrero, E and Palomares, E 2016 Carbon quantum dots as new hole transport material for perovskite solar cells *Synthetic Met.* **222** 17.

[10] Fujimoto, K, Oku, T, Akiyama, T and Suzuki, A 2013 Fabrication and characterization of copper oxide-zinc oxide solar cells prepared by electrodeposition *J. Phys. Conf. Ser.* 433 012024

[11] Pourrahimi, AM, Liu, D, Ström, V, Hedenqvist, MS, Olsson, RT and Gedde, UW 2015 Heat treatment of ZnO nanoparticles: new methods to achieve high-purity nanoparticles for high-voltage applications *J. Mater. Chem. A* **3** 17190

IOP Conf. Series: Materials Science and Engineering **625** (2019) 012018 doi:10.1088/1757-899X/625/1/012018

Nanostructured zinc oxide films grown via microwave-assisted H_2O oxidation for perovskite solar cells

C M Pelicano[1] and H Yanagi[2]

[1]Graduate School of Materials Science, Nara Institute of Science and Technology 8916-5 Takayama, Ikoma, Nara 630-0192, Japan
[2] Graduate School of Science and Technology, Nara Institute of Science and Technology, 8916-5 Takayama, Ikoma, Nara 630-0192, Japan

E-mail: christian.pelicano.ce0@ms.naist.jp, christian.mark.o.pelicano@gmail.com

Abstract. ZnO is well-known to be one of the most significant metal oxides that can be used as an electron-transporting layer (ETL) in solar cells. Hence, developing efficient and low temperature protocols have been the spotlight of numerous studies to form nanostructured ZnO films. Herein, the growth of nanostructured ZnO films via low-temperature microwave-assisted H_2O oxidation is presented. ZnO nanorods developed after 30 min of microwave irradiation. Then the nanorods evolved to nanoparticles in longer irradiation time. Raman scattering spectra of these nanostructures displayed intense E_2-mode peaks confirming their high-quality wurtzite crystal structure. All the samples have an average transmittance of ~ 80%, which demonstrates their potential as ETL for solar cells. Our results highlight the potential of the integration of microwave technology with H_2O oxidation as a fast, low temperature and sustainable process of developing ZnO nanomaterials that can be used for flexible ZnO-based electronics.

1. Introduction

The rapid improvement in efficiency of organic-inorganic hybrid lead halide perovskite solar cells (PSCs) over the past decade makes it the frontrunner alternative to Si solar cells [1]. Generally, a PSC has a device configuration of glass/ITO or FTO/electron-transporting layer (ETL)/perovskite/hole transporting layer/metal electrode [2]. Several reports are now focusing on the development of novel ETL materials to have faster electron extraction, reduced carrier recombination and ultimately, to achieve higher performance [3]. Among known metal oxides, ZnO is seen to be a viable ETL candidate owing to its low temperature processability, high transparency, and most importantly, higher mobility than TiO_2 [3-8]. In our previous work, we presented an innovative strategy to form nanostructured ZnO films through low-temperature H_2O oxidation of Zn thin films. These nanostructured ZnO films were subsequently employed as ETL for PSCs [9]. However, H_2O oxidation using a conventional hot-plate heating is time consuming and inefficient leading to non-uniform reactions.

In this study, we propose the integration of microwave technology with H_2O oxidation to fabricate novel ZnO nanostructures. Microwave heating is a promising synthesis technique for nanostructured materials due to its homogeneous volumetric heating, high reaction rate and energy consumption efficiency as compared with conventional heating methods [10]. Several studies have already reported microwave-assisted growth of ZnO nanostructures. However, they still involve organic or complexing agents which serve as impurities and are costly which hinder their commercialization [11, 12]. Thus, we propose the microwave-assisted H_2O oxidation as a facile, low cost, environmental-friendly and ultrafast process to synthesize nanostructured ZnO films.

2. Experimental section

ITO/glass substrates were cleaned with detergent, ultrapure water, acetone and methanol for 15 min each using ultrasonicator. After drying, the substrates were treated with UV/O_3 for 10 min. Prior to Zn thin film deposition, ZnO nanoparticles were spin-coated onto the substrates at 5000 rpm for 30s, which serves as a seed layer. Then, the films were annealed at 200 °C for 30 min. Zn thin films were then thermally evaporated on top of the ZnO seed layer at a base pressure of $4x10^{-4}$ Pa using a vaccum evaporator (JEOL JEE-400). The Zn thin films were subsequently immersed in a beaker filled with pure H_2O and placed inside the microwave oven. The samples were irradiated with a constant microwave power of 500 W in varying irradiation time of 30-120 min. The surface morphologies of the samples were observed using a scanning electron microscope (SEM, SU6600). Raman scattering and optical transmission spectra were measured using a Raman (JASCO NRS-4100) and a UV-Vis spectrophotometer (JASCO V-530), respectively.

3. Results and discussion

Figure 1 shows that ZnO nanorods with uneven growth length were produced even only after 30 min of microwave exposure of Zn thin films. Based on our previous work, the vacuum-deposited Zn thin films have platelet-like nanostructures. Thus, the formation of nanorods from nanoplatelets can be explained through the corrosion of Zn thin films in the presence of H_2O and dissolved oxygen in our system. As the Zn thin film is oxidized, the reduction of H2O and dissolved oxygen release OH- ions that in turn will react with Zn^{2+} ions to form ZnO crystals [9].

Figure 1. SEM images of nanostructured ZnO films formed after (a) 0.5, (b) 1, (c) 1.5 and (d) 2 h of microwave-assisted H_2O oxidation.

Moreover, uniform nanorods developed after 1 h having similar architecture as conventional H_2O-oxidized ZnO nanorods formed after 8 h [9]. The increase of irradiation time to 90 min triggered the formation of smaller nanorods. Notice also that the overall film became smoother. This phenomenon is possibly caused by the dissolution and recrystallization of ZnO in between the nanorods. Finally,

IOP Conf. Series: Materials Science and Engineering **625** (2019) 012018 doi:10.1088/1757-899X/625/1/012018

longer irradiation time of 2 h led to the development of nanoparticles which are located on top of the nanorods. The room temperature Raman scattering spectra of the samples under excitation of a 532 nm laser line are shown in Figure 2. The two most intense and sharp peaks centred at 99 cm^{-1} and 438 cm^{-1} can be assigned to the E_2-low and E_2-high modes, which are typical to the wurtzite phase of ZnO [13]. Additionally, this result validates that the nanostructured films have highly-crystalline structures.

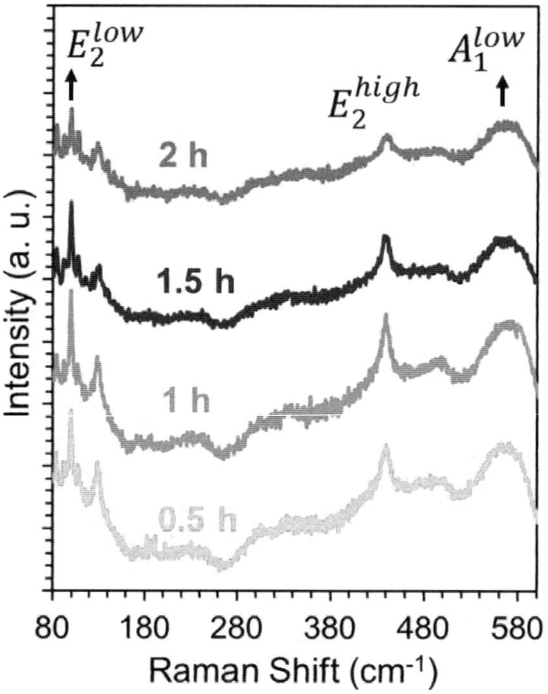

Figure 2. Raman scattering spectra of nanostructured ZnO films formed after microwave-assisted H_2O oxidation.

The solar cell device performance also strongly depends on the transparency of ETL; thus, the optical transmission spectra of the samples were measured, as presented in Figure 3. It can be seen that all the samples are highly transparent (~ 80%) in the range of 500-900 nm and the absorption edge is at about 380 nm. This excellent optical compatibility renders ZnO nanostructures prepared by the microwave-assisted H_2O oxidation as a promising electron-transporting material in perovskite solar cells.

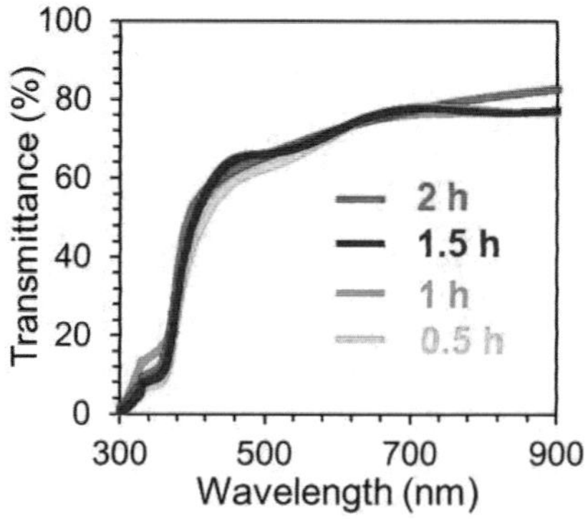

Figure 3. Optical transmission spectra of nanostructured ZnO films formed after microwave-assisted H_2O oxidation.

4. Conclusion

In summary, the fabrication of different novel ZnO nanostructures via microwave-assisted H_2O oxidation method is demonstrated. The faster rate of reaction using microwaves resulted in the morphological evolution of nanostructured films in a much shorter period of time compared with the conventional H_2O oxidation technique. Lastly, all the films are highly-crystalline and transparent. Now, we are planning to utilize these films as ETL for PSCs.

Acknowledgments

This work has been financially supported by NAIST Foundation Research Grant. The authors also gratefully acknowledge Nara Institute of Science and Technology for the experimental support.

References

[1] Park N G, 2013 *J. Phys. Chem. Lett.* **4** 2423
[2] Stranks S D and Snaith H J 2015 *Nat. Nanotechnol.* **10** 391
[3] Pelicano C M and Yanagi H 2018 *J. Energy. Chem.*, **27** 455
[4] Garcia V J, Pelicano C M and Yanagi H 2018 *Thin Solid Films* **662** 70
[5] Pelicano C M and Yanagi H 2019 *Appl. Surf. Sci.***467-468** 932
[6] Balela M D, Pelicano C M, Ty J and Yanagi H 2017 *Opt. Quant. Electron.* **49** **(1)** 3
[7] Pelicano C M, Magdaluyo E and Ishizumi A 2016 *MATEC Web of Conferences* 43, 02001
[8] Pelicano C M and Yanagi H 2017 *J. Mater. Chem. C* **5** 8059
[9] Pelicano C M and Yanagi H 2019 *J. Mater. Chem. C* DOI: 10.1039/C9TC00401G
[10] Nuchter M, Ondruschka B, Bonrath W and Gum A 2004 *Green Chem.* **6** 128
[11] Majithia R, Speich J and Meissner K E 2013 *Materials* **6** 2497
[12] Ridha N J, Umar A A, Alosfur F, Jumali M H and Salleh M M 2013 *J Nanosci Nanotechnol.* **13(4)** 2667
[13] Musa I, Qamhieh N and Mahmoud S 2017 *Results Phys.* **7** 3552

IOP Conf. Series: Materials Science and Engineering **625** (2019) 012019 doi:10.1088/1757-899X/625/1/012019

Using non-continuous gold film as the seed layer for developing zinc oxide nanowire array

P-K Chiu[1], C-W Lin[2], D-R Huang[2], C-T Yang[3] and D Chiang[1,*]

[1] Instrument Technology Research Center, National Applied Research Laboratories, Hsinchu 30076, Taiwan
[2] College of Photonics, National Chiao Tung University, Hsinchu 30010, Taiwan
[3] Material and Chemical Research Laboratories, Industrial Technology Research Institute, Hsinchu 31057, Taiwan

[*]E-mail: dony@itrc.narl.org.tw

Abstract. This study aims to discuss the grown morphology and current generation of ZnO nanowire array developed from non-continuous Au thin film as the seed and compare the results with the nanowire array from ZnO thin film as the seed layer. The zinc nitrate hexahydrate and hexamethylenetetramine used as the reaction solution and the hydrothermal process with process temperature less than 95°C are applied. It is revealed that ZnO nanowires developed with non-continuous Au film as the seed appear sparser density, present better crystallization direction (002) and higher collimation. The effect of concentration of reaction solution on the diameter, length and distribution of nanowires is reported. The electron back scattering diffraction technique confirms each grown wire is a single crystal from the bottom to top. The ZnO nanowires with piezoelectric property are measured with the conductive atomic force microscope system. The result shows that single ZnO nanowire could generate more than 800 pA current when being bent.

1. Introduction

Tian et al. opened a new era and synthesized zinc oxide nano-structure on a substrate with seed in 2002 [1]. In the same year, Boyle et al. in UK synthesized Zinc Oxide micro rods with a two-phase process [2]. The zinc oxide seed layer was first prepared at the first phase and zinc oxide micro rods were developed at the second phase. Yang et al. further developed highly aligned ZnO nanowires on the silicon substrate with the two-phase technology in 2004 [3]. Wang and Song utilized the characteristics of aligned ZnO nanowires and successfully converted mechanical energy into electric energy under nanoscale [4]. The thick and dense ZnO nanowire arrays with random preferable orientation were documented in the references [1-4].

Spin coating [5] and ZnO sputtering were used in this laboratory for preparing seed layer and successfully synthesized ZnO nanowires with distinct density and appearance by using aqueous solution with different concentrations. The piezoelectric test was further preceded. Nonetheless, it was discovered that ZnO nanowires developed with low concentration appeared to have worse collimation and that the nanowires could not easily generate power in the entangled and less alignment condition. Under high concentration, the developed nanowires presented excellent collimation, but excessive concentration resulted in difficulty deformation because of the intimate rod contact. The measured voltage/current characteristics therefore were not satisfactory [5]. The purpose of the study is to generate the nanowire arrays with controllable concentration and wire density of the nanowire array as a proposal to introduce more electric current. The gold film was used as the seed layer at the first phase in order to develop aligned ZnO nanowires with lower density was first proposed by Wang at al. but no electric performance was measured [6]. In order to enhance the piezoelectric efficiency, Lee et al. [7] produced a device using MEMS technology by covering ZnO nanowires arrays with P-type P3HT material and successfully applied ZnO nanowires to piezoelectric element to enhance the piezoelectricity.

2. Experiment method and step

The ZnO nanowire arrays were prepared by a two-phase process. The seed layers were coated on the silicon substrate using the sputtering system and the wire arrays were developed in the solutions with different concentrations using the hydrothermal technology. The silicon wafers were cleaned in the ultrasonic baths by sequential dipping and rinsing in the acetone, DI water, isopropyl alcohol, DI water, each step for 10 minutes to remove the dusts and grease in the surface and were dried by the blown N_2. The gold film of 10 nm thickness was deposited on substrate as seed layer using the DC sputtering system with working pressure of 5.7×10^{-2} torr and deposition rate of approximate 0.8 Å/s. The ZnO film of 10 nm thickness was prepared using RF sputtering system with simultaneous introduction of 25 sccm argon flow rate and 20 sccm oxygen flow rate. The reaction solutions, $Zn(NO_3)_2 \cdot 6H_2O$ and $C_6H_{12}N_4$ (HMTA) with the different concentrations, were prepared separately and then mixed and poured into a Teflon container to become a solution of predetermined concentration. The Teflon container with solution was sealed and maintained in a furnace with set temperature of 90°C for the determined time.

The morphology of the obtained ZnO nanowire array was examined by a scanned electron microscope. Microstructural analysis was performed using the X-ray diffractometry with a Cu-Kα ray. The crystal identification of single rod was observed by electron backscattering diffractometry. The bending induced current of the prepared rods was measured using a conductive atomic force microscope with a metal Pt-coated probe tip.

3. Result and discussion

3.1. Comparison of nanowire arrays developed with Au and zinc oxide as seed layers

The initial seed layer exhibits great influence on the development of ZnO nanowire arrays with hydrothermal method. The reaction solution with the concentration 0.07 M, is then added for developing ZnO nanowires. The reaction temperature is 90°C and the development time is 5 hrs. Figures 1a and 1b are the top surface SEM of nanowires developing on Au seed layer and ZnO seed layer, respectively. The ZnO nanowires present hexagonal prism on the top end of rod, conforming to ZnO wurtzite hexagonal structure. The nanowires developed on the ZnO seed layer are tighter, and the initially developed nanowires appear to be interlocked; besides, merely initially developed aligned

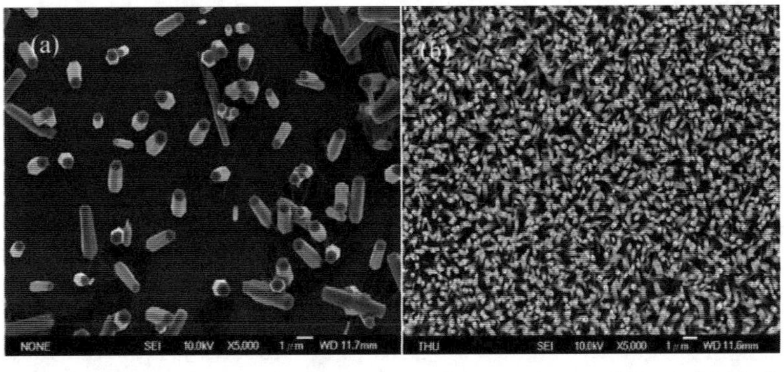

Figure 1. Comparison of ZnO arrays developed with (a) Au as the seed layer (b) ZnO as the seed layer.

nanowires could develop upwards. The ZnO nanowires show large differences in density; ZnO nanowires developed from Au seed appear more spaced, while those developed from ZnO seed are tighter. The bottom substrate cannot be viewed from the top of the array developed on ZnO seed layer substrate.

Figure 2 shows the X-ray diffraction pattern of ZnO nanowires developed from Au seed layer and ZnO seed layer. It is obvious that ZnO nanowires developed from both seed layers grow along with the preferred orientation (002) crystal plane. The crystallization direction of nanowires developed from Au seed is more consistent, while those developed from ZnO seed layer would exist in (102) and (110) directions with inconsistent crystallization direction.

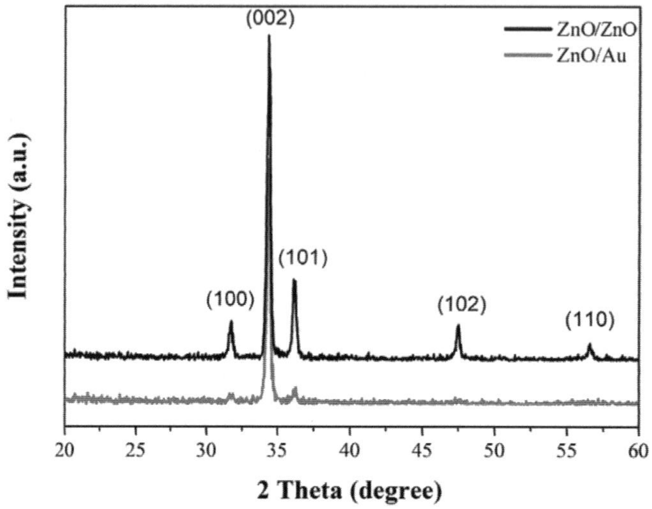

Figure 2. Comparison of XRD of nano arrays developed with ZnO (black) and Au (red).

3.2. Appearance distribution of ZnO nano arrays developed on Au seed layer with different concentration

The experiment is developed under $Zn(NO_3)_2 \cdot 6H_2O$ and $C_6H_{12}N_4$ (HMTA) with the same molar volume concentration for 5 hours, and the effect of concentration being 0.04 M, 0.07 M, and 0.1 M on ZnO nanowires is discussed. Figure 3a shows ZnO nanowires developed under the concentration 0.04 M, revealing that the diameter mainly grows between 0.4 μm~0.6 μm, with the average diameter about 0.45 μm and the nanowire density of 0.071 (1/μm^2). Figure 3b displays ZnO nanowires developed under the concentration 0.07 M; the nanowire diameter appears to be 0.6 μm~1.0 μm, with the average diameter 0.87 μm and the nanowire density 0.129 (1/μm^2). Figure 3c depicts ZnO nanowires developed under the concentration 0.1 M; the nanowire diameter is roughly 0.7 μm~1.0 μm, with the average diameter 0.73 μm and the nanowire density 0.216 (1/μm^2). The development density of ZnO

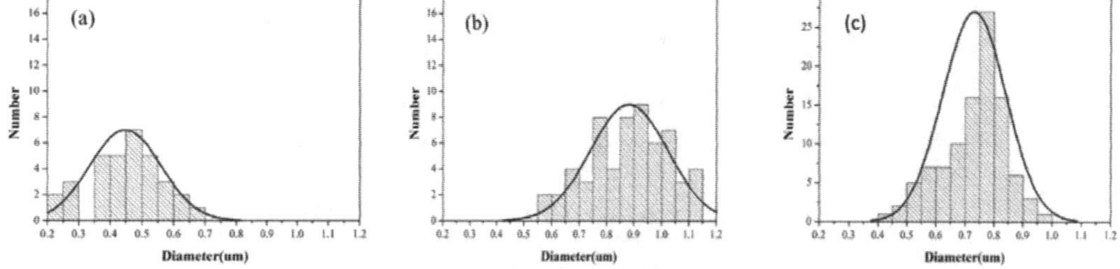

Figure 3. Comparison of nanowire diameters of arrays developed with (a) concentration 0.04 M, (b) concentration 0.07 M, and (c) concentration 0.1 M.

nanowires increases with increasing concentration mainly because the higher concentration of Zn ions in the solution will increase the number of precursors. The mobility of precursors in the reaction solution therefore is enhanced and rapidly creates higher density and larger diameter of nanowires.

IOP Conf. Series: Materials Science and Engineering **625** (2019) 012019 doi:10.1088/1757-899X/625/1/012019

3.3 Single rod was evaluated by electron backscattering diffractometry

The electron backscattering diffractometry can apply for the microstructure observation and to identify the detailed crystallographic analysis on a rod. The side view, band contrast, phase and element map are presented in Figure 4 for the ZnO rod developed on the Au seed with the development solution of 0.1M concentration. It can be seen that every single wire is a single crystal with the same crystal orientation from the bottom to the top observed by the band contrast and phase mode. The rod consists of elements Zn and O, confirmed by the EDS element analysis attached to SEM.

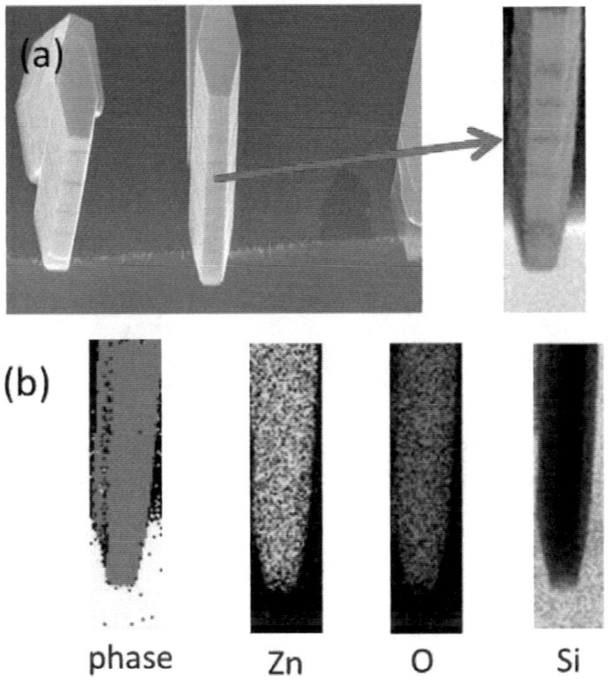

Figure 4. (a) side view and band contrast of a nanorod (b) the phase and individual element map

3.4 Electricity generation characteristics of single nanowire measured with conductive atomic force microscope

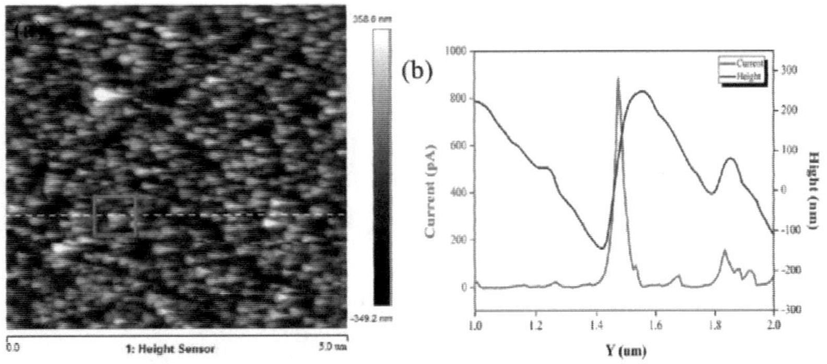

Figure 5. (a) Topography with C-AFM scanning (b) measurement of height (blue) and power generated when being knocked (red) of the correspondent red area.

Figures 5a and 5b are the topography and current diagram of ZnO nanowire arrays measured with C-AFM scanning. Figure 5a shows nanowires lining up on the substrate, and the current signals of the correspondent knocked nanowires are shown in Figure 5b; the maximum could reach 800 pA. Figure 5b is the enlargement of the red selected area in Figure 5a.

4. Conclusion

The morphology of the ZnO nanowire arrays developed from the Au and ZnO seed layers of similar thickness is compared. It is found that the density of wire arrays shows significant difference and that the appearance of the array developed from ZnO seed is more interlocked and tighter compared to that from the Au seed. The preferable orientation (002) of the array developed on Au seed is perpendicular to the substrate. The diameter of the wires increases approximately with high concentration of Zn ion precursors in solution. The length of the wires increases with development time but is not correlated with the solution concentration. It is confirmed that every whole wire grown from the solution is a single crystal. The maximum current generated from a bent wire can reach 800 pA with the current geometric feature.

Acknowledgement

The authors thank the financial support from the Ministry of Science and Technology, Taiwan.

References

[1] Tian Z R, James A, Voigt J A, Liu J, Mckenzie B and Mcdermott M J 2002 *J. Am. Chem. Soc.* **124** **(44)** 12954
[2] Boyle D S, Govender K and O'Brien P 2002 *Chem. Commun.* 80
[3] Yang X, Shao C, Guan H, Li X and Gong J 2004 *Inorganic Chem. Commun.***7(2)** 176
[4] Wang Z L and Song J 2006 *Science* **312**(5771) 242
[5] Jhong Y-H, Yang C-T, Chang L-K, Huang D-R and Chiang D 2016 *Int. J. Mater. Res.* **107**(8) 704
[6] Xu S, Lao C, Weintraub B and Wang Z L 2008 *J. Mater. Res.* 23(8) 2072
[7] Lee K Y, Bae J, Kim S M, Lee J-H, Yoon G C, Gupta M K, Kim S, Kim H, Park J and Kim S-W 2014 *Nano Energy* **8** 165

IOP Conf. Series: Materials Science and Engineering **625** (2019) 012020 doi:10.1088/1757-899X/625/1/012020

Photo-oxidation of an organosulfur for photo-charging of lithium-ion batteries

Hiroyuki Shingu[1], Hiroshi Uemachi[2] and Akihiko Fujiwara[1,*]

[1] School of Science and Technology, Kwansei Gakuin University, 2-1 Gakuen, Sanda, Hyogo 669-1337, Japan
[2] Polythione Co. Ltd., 2-13 Asahidai, Nomi, Ishikawa 923-1211, Japan

Email: akihiko.fujiwara@kwansei.ac.jp

Abstract. Polymers of organosulfur with S-S bonds are potential cathode materials for high-capacity lithium-ion batteries, and the photo-oxidation of organosulfurs can be applied in the future to the photo-charging of batteries. Herein, we present the photo-oxidation of dithiobiuret (DTB), a simple organosulfur molecule with two neighbouring tautomeric thiol/thione groups (-SH/>C=S) that can form a disulfide (S-S) bond by reversible intramolecular oxidative cyclization. Aqueous solutions of DTB with different pH values were irradiated by visible light from a halogen lamp. Photo-oxidation was confirmed by ultraviolet (UV) absorption spectroscopy, on comparison of the UV absorption spectra of DTB and chemically oxidized DTB. The rate of photo-oxidation depended on the pH of the DTB solution, and the highest photo-oxidation rate of ~ 2.8×10^{-3} $min.^{-1}$ was observed at a pH of ~ 3.1.

1. Introduction

Lithium-ion batteries (LIBs) constitute a class of the most promising energy storage devices due to their high capacity. For application in large-scale stationary energy storage, as well as for electric vehicles, further improvements in terms of their capacity and lifetime are required. Since the capacity (100 – 200 mAh/g) of the cathodes in commercial LIBs is much less than that of the anode (graphite, 372 mAh/g), development of cathode materials is a crucial issue. In this context, sulfur-containing materials can be considered as cathode materials with higher capacity, due to the small atomic mass and wide range of valence states of sulfur. In fact, many organosulfur compounds have been investigated [1-10] in addition to the sodium-sulfur (NaS) battery for stationary applications. Among organosulfur compounds, 2,5-dimercapto-1,3,4-thiadiazole (DMcT) showed high capacity, > 300 mAh/g [5,6,8-10].

The origin of the high capacity of organosulfur materials lies in using the formation/scission of the disulfide (S-S) bond as the charge/discharge processes [8]. However, many issues need to be resolved for the application of sulfur-containing materials to consumer products, such as slow redox rate at room temperature [8,9], lack of electrical conductivity, and dissolution into the electrolyte solution during the discharge (reduction) process [10]. Uemachi *et al.*[11,12] designed and synthesized a type of organosulfur polymer with S-S bonds in the side group to address these concerns—polymers with the dithiobiuret (DTB) unit incorporated into the polymer backbone. The DTB unit possesses two neighboring tautomeric thiol/thione groups (-SH/>C=S) that can form an S-S bond by reversible intramolecular oxidative cyclization [11,12]. The developed materials show high electrical conductivity and low solubility in the electrolyte owing to their polymeric form. Moreover, they have potential for exhibiting high redox rates because the sulfur in the tautomeric thiol/thione groups can react with the neighboring sulfur in the same molecule. It is well known that the scission and formation of S-S bonds through redox reactions can be controlled by photo-irradiation [13-15]. Formation of the S-S bond by photo-oxidation, therefore, can be used for photo-charging the cathode of a lithium-ion battery with the abovementioned organosulfur polymers (DTB copolymers).

Recently, a highly efficient photo-charging lithium-ion battery [16] was reported based on a combination of solar cells and rechargeable batteries [17]. However, this kind of system has disadvantages such as bulkiness and complexity due to the combination of charging (solar cell) and storage (rechargeable battery) parts in the system. Direct photo-charging based on photoelectrochemical [18,19], ferroelectric [20], and other effects can solve these issues. Even in these cases, the efficiency of direct photo-charging batteries still needs to improve for practical applications [21-25]. Direct charging of a high-capacity electrode-material by photo-irradiation is, thus, the need of the hour. Photo-oxidation of the high-capacity cathode of DTB copolymers is promising, although a systematic study of these materials has not been reported so far. In this study, we investigated the photoreactions of DTB, a model monomer material of DTB copolymers for high-capacity LIBs, and succeeded in photo-oxidation (formation of the S-S bond) of DTB by irradiation of visible light as a proof of principle (Figure 1).

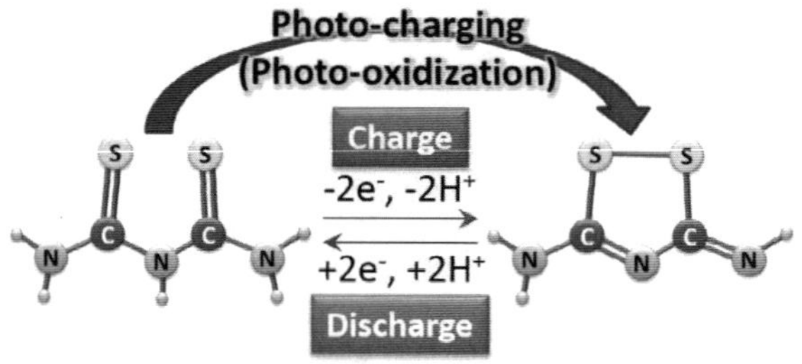

Figure 1. Schematic representation of redox reaction of DTB and the concept of photo-oxidation.

2. Experiments

DTB (CAS:541-53-7) was selected as a model sample, because it is one of the simplest organosulfur molecules. This small molecule is more appropriate than polymers, for light irradiation and UV absorption spectroscopy in aqueous solutions with a wide variation of pH. For photo-oxidation and UV absorption spectroscopy, 0.01 mM DTB was prepared in 200 ml of water. High-quality hydrochloric acid was used to control the pH of the solution. The pH of the solution was measured by a pH meter (Horiba D-75LAV).

A small volume of the sample solution (4 ml) was transferred to a quartz cuvette, and irradiated by visible light using a halogen lamp (Kenko Tokina Co., Ltd. KTS-150RSV), as shown in the inset of Figure 2. The incident light intensity at the sample position in this work was about 50 mW (~ 300 W/m^2). UV absorption spectroscopy measurements were carried out using a system from ALS instruments (SEC2000-UV/VIS). Absorbance A was calculated by

$$A = -\lg\left(\frac{I}{I_0}\right), \quad (1)$$

where I and I_0 are the intensities of the transmitted light and incident light, respectively. In this work, the spectrum of pure water in a cuvette was taken as I_0 in order to subtract the large contribution of absorption by the cuvette and water. Temporal evolution of the UV absorption spectra by visible light irradiation was measured at the irradiation time (t_i) from 0 to more than 30000 min. When the aqueous solution of DTB is not connected to an electrochemical circuit (denoted as "open"), an effect of charging-up of solution may inhibit the oxidation reactions. In order to reduce such the effect, we also conducted the measurements under zero-voltage-bias application to the solution using an electrochemical circuit with a potentiostat from ALS instruments (DY2323) for removal of charging effect (denoted as "0 V bias"). For control experiments, chemically oxidized DTB was prepared. The chemical oxidation of DTB, i.e., formation of S-S bond in DTB, was confirmed by synchrotron radiation (SR) x-ray diffraction and SR x-ray absorption fine structure (XAFS) at the BL02B1 and the BL27SU beamlines of SPring-8, respectively [26,27].

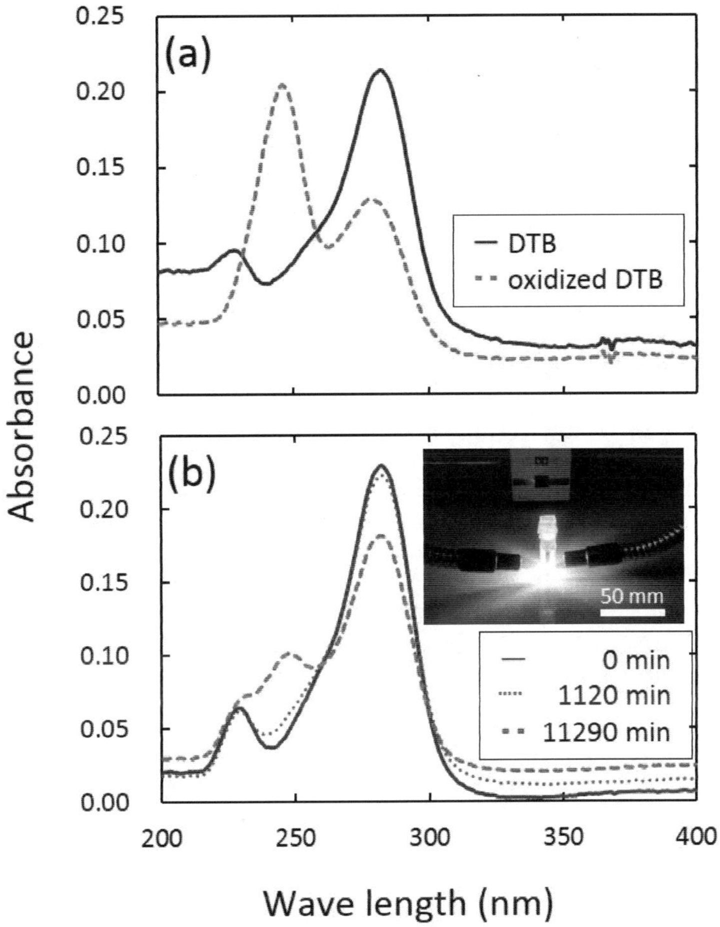

Figure 2. (a) UV absorption spectra of DTB (blue solid curve) and chemically oxidized DTB (red dashed curve). (b) The UV absorption spectra of DTB at t_i = 0 min. (blue solid curve), 1120 min. (green dotted curve), and 11290 min. (red dashed curve), in aqueous solution at a pH of 4.5. Inset shows the setup for visible light irradiation by halogen lamp. Difference in the baseline of the three spectra in (b) may originate from the change in the background of the system during the long measurement time.

3. Results and discussion

The UV absorption spectra of DTB and chemically oxidized DTB [26,27] are shown in Figure 2(a), and a clear difference was observed. The spectrum of DTB can be represented by a linear combination of three Gaussian functions with peaks at 226, 256, and 282 nm, while that of chemically oxidized DTB can be represented by a linear combination of two Gaussian functions with peaks at 245 and 279 nm. A decrease in the peak intensity around 300 nm (282 nm in this work) [14,15] and the appearance of a peak at 245 nm are an indication of the oxidation of DTB. Figure 2(b) shows typical examples of the UV absorption spectra of an aqueous solution of DTB with a pH of 4.5, at t_i = 0, 1120 and 11290 min. for visualizing the temporal evolution of UV spectra. As t_i increases, the typical peak of the oxidized DTB at 245 nm increases in intensity. On the other hand, the main peak at 282 nm originating from DTB decreases.

The contributions of DTB and oxidized DTB to the UV absorption spectra at various t_i were estimated by a linear combination of the spectra of DTB and chemically oxidized DTB. The intensities were normalized based on the intensity of DTB spectrum at $t_i = 0$ and the relative intensity of chemically oxidized DTB to DTB in Figure 2(a). Figure 3 shows the dependence of the normalized UV absorption intensities of DTB and oxidized DTB upon t_i with/without light irradiation and with/without removal of charging effect at a pH of 6.4. The normalized intensity I of DTB/oxidized DTB decreases/increases upon t_i, and can be expressed as

$$I = \exp\left(-\frac{t_i}{\tau}\right), \quad (2)$$

$$I = 1 - \exp\left(-\frac{t_i}{\tau}\right), \quad (3)$$

with the time constants τ, respectively. The τ which is inversely proportional to the reaction rate strongly depends on the experimental conditions: with/without light irradiation and with/without removal of charging effect. The τ values for data shown in Figure 3 are summarized in Table 1. It should be noticed that an increase in the intensity of the oxidized DTB spectrum corresponds to a decrease in the intensity of DTB spectrum at each pH, indicating that DTB is converted to oxidized DTB with an S-S bond through photo-oxidation. From these results, it can be concluded that the photo-oxidation of DTB efficiently occurs under light irradiation at 0 V bias application for removal of charging effect.

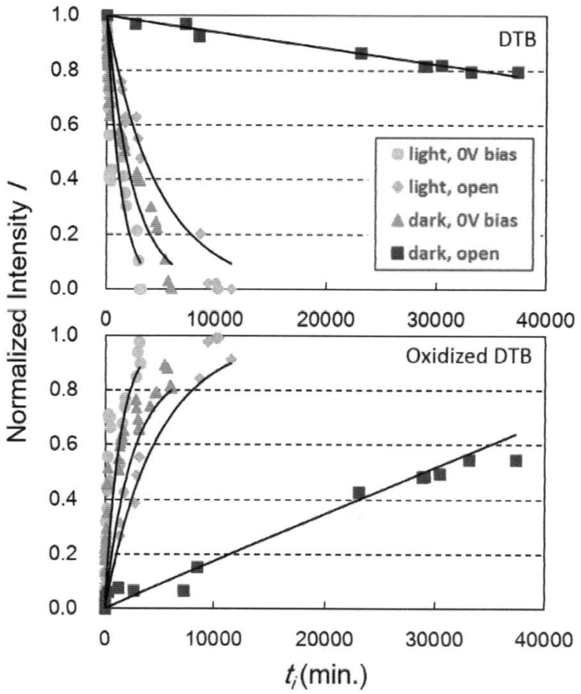

Figure 3. Visible light irradiation time (t_i) dependence of normalized UV absorption intensity of (a) DTB and (b) oxidized DTB with/without light irradiation (denoted as light/dark) and with/without removal of charging effect (denoted as 0 V bias/open).

Table 1. Photo-oxidation time constant τ for various experimental conditions at a pH of 6.4.

Experimental conditions		τ (min)
With light irradiation (light)	With removal of charging effect (0 V bias)	1.4×10^3
	Without removal of charging effect (open)	4.9×10^3
Without light irradiation (dark)	With removal of charging effect (0 V bias)	2.6×10^3
	Without removal of charging effect (open)	5.3×10^5

Figure 4 shows a relation between the photo-oxidation rate ($1/\tau$) of DTB and the pH value of the solution under light irradiation at 0 V bias application for removal of charging effect. The most efficient photo-oxidation rate, ~ 2.8×10^{-3} min.$^{-1}$, was achieved at the condition of a pH of ~ 3.1.

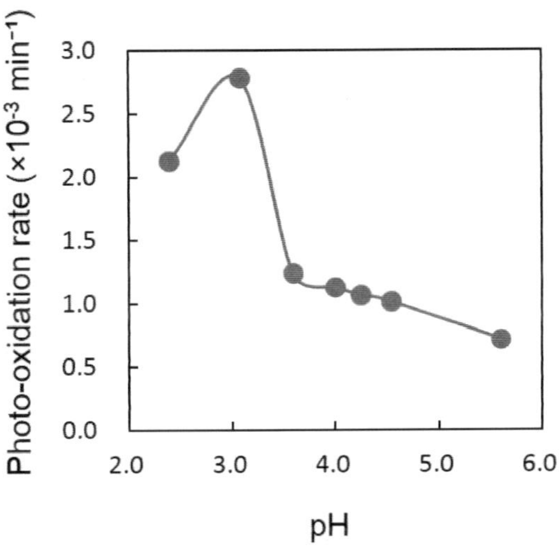

Figure 4. Photo-oxidation rate from DTB to oxidized DTB at 0 V bias application for removal of charging effect for various pH values.

4. Conclusions

We have succeeded in the photo-oxidation of dithiobiuret (DTB), a simple organosulfur molecule with two neighboring tautomeric thiol/thione groups (-SH/>C=S) that can form a disulfide (S-S) bond by reversible intramolecular oxidative cyclization. The abundance ratio of oxidized DTB increased with increasing visible light irradiation time. It was found that the photo-oxidation rate depends strongly on the experimental conditions, such as pH of the DTB aqueous solution and removal of charging effect. The highest photo-oxidation rate, 2.8×10^{-3} min.$^{-1}$, was achieved at the condition of pH of 3.1 and 0 V bias application for removal of charging effect. For the application of photo-oxidation of DTB to photo-charging of lithium-ion batteries, further improvement of photo-oxidation rate and the demonstration of photo-oxidation of DTB copolymers in the cathode of LIBs are challenges for future research.

Acknowledgements
The authors acknowledge Dr. K. Sugimoto and Dr. Y. Tamenori of Japan Synchrotron Radiation Research Institute for the synchrotron radiation experiments and fruitful discussions. The synchrotron radiation experiments were carried out with the approval of the Japan Synchrotron Radiation Research Institute (Proposals 2011B1882, 2012A1652, 2012A1653, 2012B1811). This work was partially supported by JSPS KAKENHI Grant Number JP18K05306.

References
[1] Visco S J, Mailhe C C, De Jonghe L C and Armand M B 1989 J. Electrochem. Soc. **136** 661
[2] Liu M, Visco S J and De Jonghe L C 1989 J. Electrochem. Soc. **136** 2570
[3] Visco S J, Liu M and De Jonghe L C 1990 J. Electrochem. Soc. **137** 1191
[4] Liu M, Visco S J and De Jonghe L C 1990 J. Electrochem. Soc. **137** 750
[5] Liu M, Visco S J and De Jonghe L C 1991 J. Electrochem. Soc. **138** 1891
[6] Liu M, Visco S J and De Jonghe L C 1991 J. Electrochem. Soc. **138** 1896
[7] Genie`s E M., Picart S 1995 Synth. Met. **69** 165
[8] Oyama N, Tatsuma T, Sato T, and Sotomura T 1995 Nature **373** 598
[9] Sotomura T, Uemachi H, Takeyama K, Naoi K and Oyama N 1992 Electrochim. Acta **37** 1851
[10] Naoi K, Kawase K and Inoue Y 1997 J. Electrochem. Soc. **144** L173
[11] Uemachi H, Iwasa Y and Mitani T 2000 Chem. Lett. **29** 946
[12] Uemachi H, Iwasa Y and Mitani T 2001 Electrochim. Acta **46** 2305
[13] Cremonini M A, Lunazzi L 1993 G. Placucci, J. Org. Chem. **58** 3805
[14] Bucher G, Lu C and Sander W 2005 Chem. Phys. Chem. **6** 2607
[15] Kim H J, Yoon J H and Yoon S 2010 J. Phys. Chem. A **114** 12010
[16] Xu J, Chen Y and Dai L 2015 Nat. Commun. **6** 8103
[17] Dutta P 1992 Nature **358** 621
[18] Chang K C, Heller A, Schwartz B, Menezes S and Miller B 1977 Science **196** 1097
[19] Licht S 1987 Nature **330** 148
[20] Chu B J, Zhou X, Ren K, Neese B, Lin M, Wang Q, Bauer F and Zhang Q M 2006 Science **313** 334
[21] Lo C W, Li C and Jiang H 2011 AIP Adv. **1** 042104
[22] Guo W, Xue X, Wang S, Lin C and Wang Z L 2012 Nano Lett. **12** 2520
[23] Liu P, Yang H X, Ai X P, Li G R and Gao X P 2012 Electrochem. Commun. **16** 69
[24] Yan N F, Li G R, Pan G L and Gao X P 2012 J. Electrochem. Soc. **159** A1770
[25] Yu M, Ren X, Ma L and Wu Y 2014 Nat. Commun. **5** 5111
[26] Sugimoto K, Uemachi H, Maekawa M and Fujiwara A 2013 Cryst. Growth Des. **13** 422
[27] Uemachi H, Tamenori Y, Itono T, Masuda T, Shimoda T and Fujiwara A 2018 Electrochimica Acta **281** 99

Development of Nano TiO_2 Coated Cement Surface Degradation Properties by the Photocatalytic

Pheeraphong Bunroek[1, a] and Pat Sooksaen[2,b]

[1]College of Teacher Education, Phranakhon Rajabhat University, Bangkok, Thailand
[2]Department of Materials Science and Engineering, Faculty of Engineering and Industrial Technology, Silpakorn University, Nakhon Pathom, Thailand

E-mail: [a]pheerabun@gmail.com, [b]sooksaen_p@su.ac.th

Abstract. The objectives of this study were 1)to investigate the stability of Colloid Nano-particles by using diffusion aid, 2)to study the TiO2 nanoparticles on the cement surface, and 3)to study the organic degradation behavior on the TiO2 coated cement surface by the Photocatalytic. The study of the stability of the dispersant found that Polyethyleneglycol (PEG) MW 4000 was more effective in dispersing nanoparticles than other materials. The sample size was 5x5x1 centimeters, and cemented to water/cement ratio of 0.4, and hydration cement for 28 days at room temperature. Surface was grinded and sprayed coating of TiO2 nanoparticles on the cement surface with Air Brush. Then, the SEM confirmed it. There were nanoparticles attached to the cement surface and in the porous area of the cement. Methylene blue (MB) concentrations of 0.01, 0.03, and 0.05 mg/L for 3 hours under UV light (in UV box) were measured. The TiO2 nanoparticles could decompose organic matter by the Photocatalytic. The observation of the reduced absorbance of the nano TiO2 coated cement surface was measured with UV-Vis technique. It was ensured that nano TiO2 coated cement surface was able to decompose organic substances and apply them effectively to outside walls or cementitious materials.

1. Introduction

Thailand is geographically tropical. There is a risk of exposure to microorganisms accumulated in the environment, especially the moisture cement wall. The cement wall is dirty from fungus. In this research, the researchers wanted to develop the cement surface that could eliminate the pathogenic microorganisms by the Photocatalytic [1-3]. The objectives of this study were 1) to investigate the stability of Colloid Nanoparticles by using diffusion aid, 2) to study the behavior of TiO_2 nanoparticles on the cement surface, and 3) to study the organic degradation behavior of the nanoparticles on the cement surface by the Photocatalytic.

2. Experimental

In this study, TiO_2 nanoparticles used were from China. The main crystalline structure was Anatase, with an average diameter of 21 nm and specific surface area of 50 m^2/g. Then, the researchers synthesized the colloidal TiO_2 by using 4 dispersants: 1.Polyethyleneglycol 2. Sodiumsilicate 3.Ammoniumacetate 4.Magnesiumchloride. The comparison of the dispersant concentration were at 0.5%, 1.5% and 2.5% by weight Apply the solution to the ultrasonic machine for 30 minutes to make the nanotitania particles dispersed uncluttered[4]. It was determined that the most stable concentration was applied to the specimen. Cemented samples were mixed with water/cement ratio of 0.4 and samples size is 5x5x1 cm and hydration cement for 28 days at room temperature, wash with ultrasonic

IOP Conf. Series: Materials Science and Engineering **625** (2019) 012021 doi:10.1088/1757-899X/625/1/012021

machine and place in room temperature for 3 days, and then coated with nanoparticles without painting. After that, the work was sprayed with colloidal material of nanotitania on the cement surface with air brushes on both sides of the cement surface. The confirmation of cementitious attach of nanotitania particles was observed by SEM. The photocatalytic activity test was carried out in 0.01 ml, 0.03 and 0.05 mg/l methylene blue (MB), at 3 hours under UV light (in UV box) at constant light intensity. This study observed on the decomposition of Beer Lambert's law by measuring the concentration of MB before (C_0) and after testing (C) in the UV-Box by using UV-Vis technique.

3. Results and discussion

3.1 Synthesis of colloidal TiO_2

The researchers synthesized colloidal titanium dioxide, using four dispersants: 1.Polyethyleneglycol 2.Sodiumsilicate 3.Ammoniumacetate 4.Magnesiumchloride [4,5]. The comparison of the concentrations of the four dispersants were at 0.5%, 1.5% and 2.5% by weight, which was carried out for 15 days as shown in Fig.1 (a) (b) (c) (d) and (e). Polyethyleneglycol (PEG) with a molecular weight of 4,000 had the potential to make the nanoparticles more dispersible in water than any other substance shown in Fig.1 (b). The polymer that adsorbed on the TiO_2 nanoparticle leads to steric repulsion between particles and averts the agglomeration of nanoparticles. Based on some researches, several polymers were used for steric stabilization of TiO_2 nanoparticles. Polyethylene glycol (PEG) and polyacrylamide (PAM) for steric stabilization TiO_2 sols were used by Diess et al [6]. The TiO_2 sols Coated with PEG promotes the stability of these sols at pH below 3 and concentration up to 30 $g.l^{-1}$. However, the polyacrylamide (PAM) stabilized sols of concentration close to 100 $g.l^{-1}$ at pH up to 5, which lead to long-term stability of these sols [7]. If the colloidal titanium dioxide particles were well dispersed, the nanotitania particles would spread to the cement surface. This type of coating was applied to the cement surface to test for the organic substances decomposition.

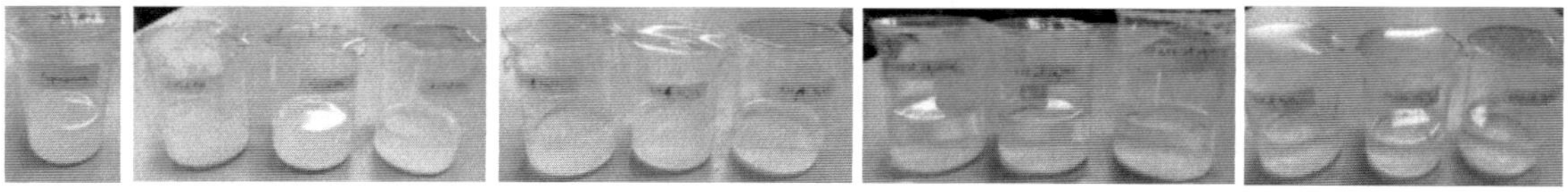

(a) (b) (c) (d) (e)

Figure 1. Colloidal titanium dioxide in days 15 (a) No dispersing agent (b) Polyethyleneglycol (c) Sodiumsilicate (d) Ammoniumacetate (e) Magnesiumchloride

3.2 Cement sample production

The sample size was 5x5x1 centimeters, and cemented to water/cement ratio of 0.4, and hydration cement for 28 days at room temperature in Fig.2 (a). Physical appearance of the sample was rough. Surface grinding was required to give smooth surface and uniform spraying of nanotitania can be assumed. Then, the cement sample was cleaned with ultrasonic machine to allow the cement powder to come out and leave it to dry for 3 days at room temperature shown in Fig.2 (b) and (c). The cement sample was sprayed coating of TiO_2 nanoparticles with PEG dispersant on the cement surface with Air Brush in Fig.2 (d)

(a) (b) (c) (d)

Figure 2. Cement Sample after 28 days (a) Cement sample before surface grinding (b) Surface grinded (c) after wash and (d) Spray coating of TiO_2 nanoparticles on the cement surface with Air Brush

3.3 Testing

3.3.1 Confirmation of cementitious attach of nanotitania particles by SEM. The nanotitania was very small at the nanometer level which could not be seen with the naked eyes or normal microscope. Nanotitania particles which attached to the surface of the cement were confirmed by using scanning electron microscope (SEM) shown in Fig.3. In Fig.3 (a) 60,000 times SEM showed that there was a nanoparticle attached to the sample surface in the circle. This confirmed that nanotitania particles attached to the sample surface in a scattering pattern. At 60,000 times SEM could not be seen within the cement porous. Fig.3 (b) at 100,000 times SEM showed that the nanoparticles embedded and distributed in the cement porous in the circle. When increasing the magnification at 300,000 times in the 200 nm range as shown in Fig. 3 (c), it was found that the nanoparticle attached to the surface and the porosity of the cement sample. To observe exactly what was found in the image, the observation of SEM was a real nanoparticle. The researchers increased in 500,000 times in the 100 nm range to measure the size of nanoparticles. Fig. 3 (d) showed that the size of nanoparticles at the surface was approximately 25 nanometers. A group of nanoparticles might be found larger than 25 nanometers. The researchers were able to confirm that the colloidal coating of colloidal titanium particles was at the nano level. Nanotitania particles were attached to the surface and porous at the surface of the cementitious specimen to be tested with MB. It was also possible that the nanoparticle-coated specimens could be used to decompose the organic material by photocatalytic test.

(a) (b) (c) (d)

Figure 3. SEM image (a) 60,000x (b) 100,000x (c) 300,000x (d) 500,000x

3.3.2 Photocatalytic activity test. Methylene blue concentrations of 0.01, 0.03 and 0.05 mg/l were used as a molecular organic reagent for the photocatalytic activity test. We applied the TiO_2 coated cement sample in MB for 3 hours under UV light (in UV box) at a constant light intensity to substitute the reaction in natural light with UV light from a UV box that was darker than the UV in the sun for reducing the study period that shown in Fig.4 (a). Uncoated cement sample soaked in MB for 3 hours to test the porosity effect to absorb MB shown in Fig.4 (b). Due to the nature of the porous cement, which might adversely affect the absorption of organic matter into the porous cement. We wanted to know the efficiency of the organic decomposition of nanoparticles attached to the cement surface. This experiment was conducted to compare the effect of immersion of cement coated with nanotitania nanoparticles in MB solution. Then, the decreasing of MB concentration in UV box was measured by Beer Lambert's Law.

(a) Nanotitania coated in UV box (b) Uncoated in laboratory

Figure 4. Cement Sample Soaked in Methylene Blue

Over the course of 3 hours, we observed the color change of Methylene Blue. The blue color of the solution was reduced from the standard solution. The absorbance was measured by UV-vis technique at concentrations of 0.01, 0.03 and 0.05 mg/l in three batches in Fig.5 (a) (b) and (c), respectively. The first was the standard solution shown in the line MB001mM_std Abs., MB003mM_std Abs. and MB005mM_std Abs. The second was the solution obtained after immersion in uncoated cement with nanotitania particles as shown in line MB001mM_c3h Abs., MB003mM_c3h Abs. and MB005mM_c3h Abs. The last solution from the cement sample was coated with nanotitania nanoparticles and immersed in MB under UV light in the UV box shown in line MB001mM_cn3h Abs. MB003mM_cn3h Abs. and MB005mM_cn3h Abs.

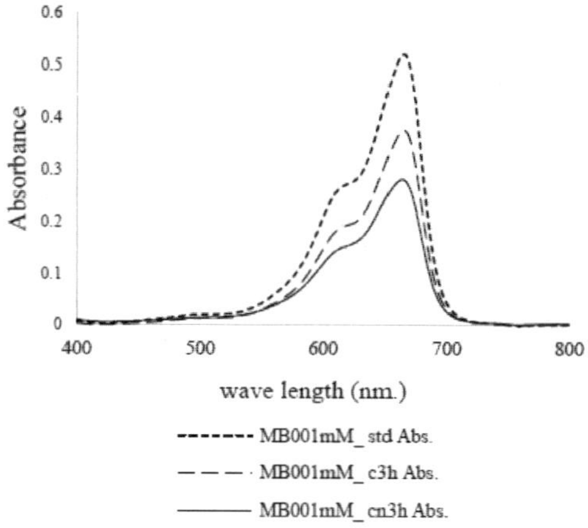

(a) Absorbance at 0.01 mg/l

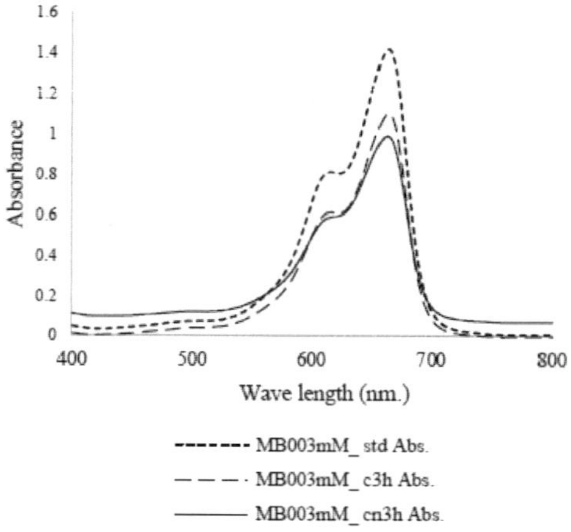

(b) Absorbance at 0.03 mg/l

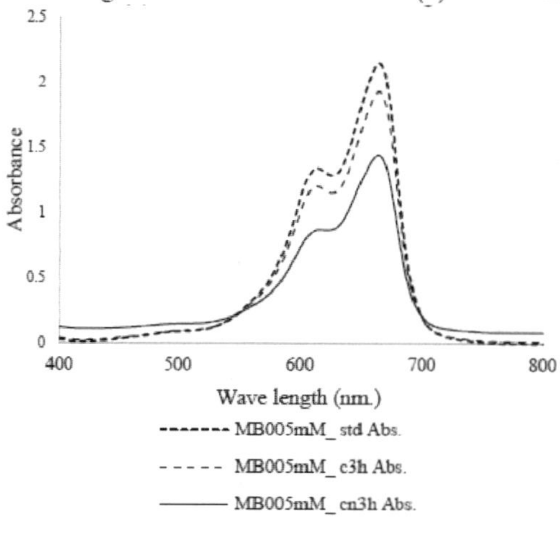

(c) Absorbance at 0.05 mg/l

Figure 5. The graph shows the absorption of the Methylene Blue solution.

Fig.5 (a) showed the absorption of the solution at 0.01 mg/l. The wavelength was 664 nm. The first was a standard solution. The absorbance was 0.52. The second was the solution obtained after immersion in uncoated cement with nanotitania particles. The absorbance was 0.37. The absorb-ance was reduced from the standard solution by about 28.8% with the effect of cement porous absorbing. The third was the solution from coated cement with nanotitania nanoparticles was immersed Methylene Blue under UV light in a UV box with an absorbance of 0.28. The absorbance decreased from the standard solution of 46.1%. The effect of the cement porosity and nanotitania nanoparticles on the cement surface was effective when exposed to light. Fig.5 (b) showed the absorption of the solution at 0.03 mg/l. The wavelength was 664 nm. The first was a standard solution. The absorbance was 1.42. The second was the solution obtained after immersion in uncoated cement with nanotitania particles. The absorbance was 1.10. The third was the solution from coated cement with nanotitania nanoparticles was immersed Methylene Blue under UV light in a UV box with an absorbance of 0.99. The absorbance was decreased from the standard deviation of 29.28%. The effect of the cement porosity and nanotitania nanoparticles on the cement surface was effective when exposed to light. Fig.5 (c) showed the absorption of the solution at 0.05 mg/l. The wavelength was 664 nm. The first was a standard solution. The absorbance was 2.15. The second was the solution obtained after immersion in uncoated cement with nanotitania particles. The absorbance was 1.98. The absorbance was reduced from the standard solution by about 9.8% with the effect of cement porous absorbing. The third was the solution from coated cement with nanotitania nanoparticles was immersed Methylene Blue under UV light in a UV box with an absorbance of 1.44. The absorbance was decreased from the standard deviation of 32.71%. The effect of the cement porosity and nanotitania nanoparticles on the cement surface was effective when exposed to light. Because of the porosity of cement, it was beneficial to absorb organic matter in this area, working with nanotitania nanoparticles coated at the surface and into the porous cement. The results were in lower absorbance, which meant more organic degradation. Even though we tested the concentration of the Methylene Blue at the increased concentrations of the first and the second sets, the cement surfaces coated with nanotitania particles still exhibited organic degradation. Under the light, the test results were based on the assumption. When using real estate in buildings, it was more effective to work together.

4. Conclusion

It was proposed to develop self-cleaning cement by light accelerator reaction. The sample cement surface was coated with colloidal titanium nanoparticles. Using four dispersing agents, it was found that the dispersant agent Polyethyleneglycol (PEG) with a molecular weight of 4,000 was more effective in dispersing nanoparticles than others. The nano-TiO_2 nanoparticles were sprayed onto the cement surface. Confirmation of nanoparticle attachment with SEM, it was found that nanoparticles were attached at the surface and porous of the sample specimens. The coated cement was tested by photocatalytic activity test for 3 hours in UV box with 0.01, 0.03 and 0.05 mg/l MB. On the cement surface at constant light intensity, nanotitania particles could decompose organic matter in light acceleration. Observation of the reduced absorbance from UV-Vis technique was able to assure that the coated cement was able to decompose the organic substance and applied it to the outside walls or cementitious materials in the sunlight effectively.

References

[1] Jun Chen, Shi-cong Kou, Chi-sun Poon. (2011). *Photocatalytic Cement-based Materials: Comparison of Nitrogen Oxides and Toluene Removal Potentials and Evaluation of Self-cleaning Performance*. Building and Environment, Vol. 46, Issue 9, P.1827–1833.

[2] Safaei-Naeini, Y., Aminzare, M., Golestani-Fard, F., Hhorasanizadeh, F., & Salahi, E. (2012, March). *Suspension Stability of Titania Nanoparticles Studies by UV-vis Spectroscopy Method*. Iranian Journal of Material Science & Engineering, Vol 9, 62-68.

[3] Weiguo Shen, Chuang Zhang, Qiu Li, Wensheng Zhang, Liu Cao, Jiayuan Ye (2015). *Preparation of Titanium Dioxide Nano-particle Modified Photocatalytic Self-cleaning Concrete*. Journal of Cleaner Production 87. P 762-765.

[4] H. Tada, O. N. (2007). *Dispersion Stability of TiO_2 Nanoparticles Covered with SiO_2 Monolayers in Water*. Journal of Colloid and Interface Science, Vol. 306, 274-280.
[5] Siti Hajar Othman, Rashid, S. A., Ghazi, T. I., & Abdullah, N. (2012). *Dispersion and Stabilization of Photocatalytic TiO_2 Nanoparticles in Aqueous Suspension for Coatings Application*. Nanomaterials, 1-10.
[6] Deiss J, Anizan P, El Hadigui S, Wecker C (1996). *Steric stability of TiO_2 nanoparticles in aqueous dispersions*. Colloids Surf A Physicochem Eng Asp 106(1). P 59-62.
[7] Kanehira K, Banzai T, Ogino C, Shimizu N, Kubota Y, Sonezaki S (2008). *Properties of TiO_2 – polyacrylic acid dispersions with potential for molecular recognition*. Colloids and Surfaces B: Biointerfaces 64(1). P 10-5.

A novel inertial sensor exploiting magnetic fluid

X Zhang[1,], X Xia[1], W Zhang[1], Z Xiang[1]

[1] State Key Laboratory for Manufacturing System Engineering, Xi'an Jiaotong University, Xi'an 710049, China

Email：zhang-xining@163.com

Abstract. Narrow available frequency range and low linearity of traditional magnetic fluid inertial sensors make them unpractical. A novel inertial sensor exploiting magnetic fluid is designed to solve these shortcomings. Through simulation and experiments, characteristics of magnetic fluid such as damping effect and levitation effect are studied. It shows that magnetic fluid has excellent damping effect which can be controlled by magnetic field. In order to improve the frequency response, both levitation and damping effect are used in sensors. Cylindrical permanent magnet is employed as the inertial mass, which can transform acceleration to the change of magnetic field. Linear hall sensor with high sensitivity is the transducer installed on the position where magnetic field gradient of cylindrical permanent magnet is maximal and has best linearity. Verification experiments of the designed sensor are conducted on vibration platform. Experimental results give that application of magnetic fluid for inertial sensor optimizes its performance greatly. Available frequency range of the designed sensor is increased by 256% after using magnetic fluid as damping material. This paper provides a reference for research on magnetic fluid and its application in inertial sensors.

1. Introduction

Magnetic fluid is a kind of intelligent material, which is a stable colloidal levitation of magnetic nanoparticles in liquid carrier [1]. Characteristics of magnetic fluid can be controlled by magnetic field. Due to special physical performance, it has been applied in different areas including aviation, machine tools, sensors and so on [2-3]. Sensors are the core of industrial reform. As a result, application of magnetic fluid in inertial sensors has attracted attention of many scholars in recent years [4-6]. Different inertial sensors have been designed and tested. Due to the restriction of structure, current sensors are mostly used under low frequency conditions (below 15 Hz) [7-8]. Although the application of magnetic fluid in inertial sensors has achieved some achievements, the existing inertial sensors has poor practicality because of large volume and low linearity. Research strategies for magnetic fluid performance are inadequate and comprehensive, so that advantages of magnetic fluid are not fully utilized in their design. In other words, it will bring bad effects if magnetic fluid is used in wrong ways. Thus, it is vital to study the magnetic fluid and realize best application in sensors.

In this paper, characteristics of magnetic fluid under the action of magnetic field has been researched. In order to make full use of the advantages of magnetic fluid, special structure of sensors is designed. Magnetic fluid is used as the main damping component of the inertial sensor. The available frequency range of the inertial sensor has been broadened greatly. In addition to this, magnetic fluid can avoid mechanical friction and improve the service life of sensors. Both simulation and experiments are conducted to verify the feasibility and effectiveness to apply magnetic fluid in inertial sensors.

IOP Conf. Series: Materials Science and Engineering **625** (2019) 012022 doi:10.1088/1757-899X/625/1/012022

2. Material and sensor design

In order to verify the feasibility of using magnetic fluid in inertial sensors and use sensors to study magnetic fluid properties in turn, a sensor with new structure is designed. In this section, material used in our research and the design of sensor will be introduced in detail.

2.1. Material

Magnetic fluid is a stable colloidal levitation of magnetic nanoparticles in liquid carrier. Magnetic nanoparticles form a rigid chain structure under the action of magnetic field, so that magnetic fluid is polarized along the direction of magnetic field. Due to the physical changes, magnetic fluid has many special characteristics. As figure 1 shows, magnetic fluid is attracted by permanent magnet evenly because of its magnetorheological effect. During the movement, the carrier liquid is forced to flow around the magnetic nanoparticle chain, resulting in an increase of the magnetic fluid viscosity. Under the action of a gradient magnetic field, the magnetic fluid can suspend an object that is denser than itself. These two characteristic are called separately as damping effect and levitation effect, which are the research focus of this paper.

Figure 1. Magnetorheological effect of magnetic fluid

The material model used in this paper is list in table 1.

Table 1. Model of magnetic fluid used in this paper.

Attributes	Value
Liquid carrier	Kerosene
Diameter of magnetic nanoparticle (nm)	1
Viscosity(Original) ($Pa \cdot s$)	1
Viscosity(under the action of magnetic of field) ($Pa \cdot s$)	1.149
Saturation magnetization (kA/m)	31.8

2.2. Sensor design

As illustrated in figure 2(a), the designed sensor has six components. The inertial mass is two identical cylindrical permanent magnets, while the spring is a single cantilever spring piece. Permanent magnets are mounted symmetrically at the end of the cantilever. Upper and lower part of the casing are connected by screws. Spring piece is fixed between these two parts of casing. Internal cavity is filled with magnetic fluid. Besides this, high sensitivity linear Hall element is mounted at the central position of upper casing. The sensor is a second-order spring vibration system. The vibration mode is shown in figure 2(b). After being added magnetic fluid, the vibration parameters changes. Magnetic fluid is completely magnetized under the action of the gradient magnetic field formed by the permanent magnet. Due to the effect of the levitation and damping effect, the system stiffness is increased by k_f and the damping coefficient is increased by c_f . Because magnetic fluid is strongly attracted by the magnet surface due to the magnetic force, it can be considered as added inertial mass m_f .

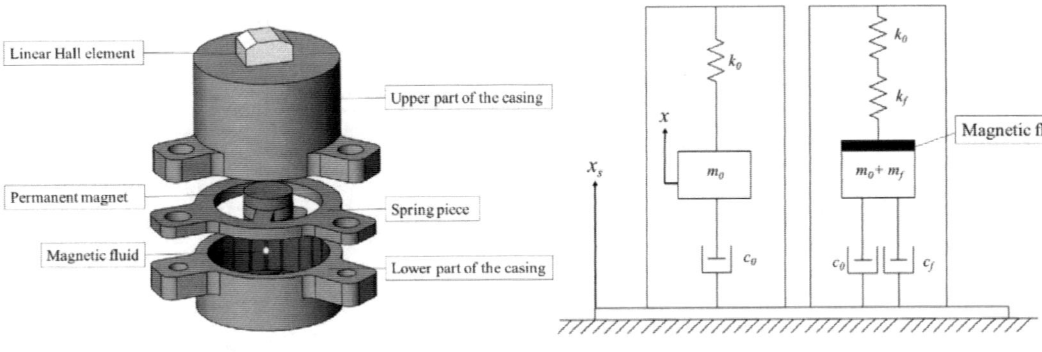

Figure 2. Sensor structure and vibration mode

The detailed information of components is shown in table 2.

Table 2. The detailed information of components.

Components	Model
Outer casing	Photosensitive resin
Permanent magnet	NdFeB35
Spring material	65Mn steel
Linear Hall element	EQ-730L

The detection principle of the inertial sensor designed in this paper is shown as follows:

- The sensor is activated by external acceleration.
- Permanent magnets run out of the balanced position because of inertial force.
- Magnetic field strength of the position where Hall element is installed changes.;
- Hall element converts magnetic field into electrical signal output.

In summary, the acceleration can be converted to the electrical signal, so as to realize the acceleration measurement. The vibration signal of sensor will be used to analyse the physical properties in the following sections.

3. Research on properties of magnetic fluid

Damping and levitation effect are main research focuses in this paper. With the designed sensor as platform, effects mentioned can be researched. Theoretical and experimental research on the damping and levitation effect will be introduced in this section.

3.1. Levitation effect

Levitation effect means that magnetic fluid can suspend an object that is denser than itself. Because the inertial mass used in the sensor is permanent magnet, second-order levitation force produced by magnetic fluid is the main research point.

3.1.1. Theoretical calculation. Firstly, theoretical analysis is conducted. The calculation result will be used to confirm simulation parameters. The calculation of the levitation force is based on a magnetic fluid micro-element. When suspending in magnetic fluid, physical force and surface stress of the magnet are balanced (see equation (1)).

$$\iint_S T_m^{'} dS = \iiint_V f_m^{'} dV \tag{1}$$

IOP Conf. Series: Materials Science and Engineering **625** (2019) 012022 doi:10.1088/1757-899X/625/1/012022

Where $T_m^{'}$ represents surface stress tensor, dV represents the volume of magnetic fluid micro-element, dS represents its surface area and $f_m^{'}$ represents physical force of per unit volume magnetic fluid.

According to the divergence theorem, equation (2) holds.

$$f_m^{'} = \nabla \cdot T_m^{'} \tag{2}$$

Assuming that the magnetic field strength is parallel to the magnetic induction, the surface stress tensor of the magnetic fluid can be obtained (see equation (3)).

$$T_m^{'} = -(p + Q_M + L_M)\mathbf{I} + \mathbf{BH} \tag{3}$$

Where $p = p(\rho_f, T)$, $Q_M = \int_0^H \mu_0 \left[\frac{\partial(vM)}{\partial v}\right]_{H,T} dH$, $L_M = \frac{1}{2}\mu_0 H^2$, p is hydrostatic pressure, T is temperature, μ_0 is vacuum permeability, M is magnetization, ρ_f is magnetic fluid density and v is magnetic fluid specific volume.

Equations (4) is the boundary conditions at the interface between magnetic fluid and permanent magnet.

$$B_n^{+} - B_n^{-} = 0, \; H_t^{+} - B_t^{-} = 0 \tag{4}$$

Where superscripts +, - respectively indicate the outer and inner surfaces of the interface.

Substituting equation (3) and equation (2) into equation (1) and taking boundary conditions into consideration, second-order levitation force can be calculated (see equation (5)).

$$F_m^{'} = \oint \left[-\left(T_M + \frac{1}{2}\mu_0 M_n^{2}\right)\right] ndS \tag{5}$$

Where $T_M = \mu_0 \int_0^H MdH$, and subscripts n represents the normal components.

According to equations (5), second-order levitation force can be calculated. However, parameters can be quantized. Simulation based on certain assumptions is the choice. Mirror method is used to carry out the simulation. The magnetic field generated by the permanent magnet is equivalent to the magnetic field generated by the mirror magnet, and the second-order levitation force of the permanent magnet is the interaction force with the mirror magnet. The residual magnetic vector of the mirrored permanent magnet is opposite to that of the original permanent magnet. The calculation formula of residual magnetic vector is shown as equations (6).

$$J_{img} = \frac{\mu_r - \mu_r^{'}}{\mu_r + \mu_r^{'}} J_0 \tag{6}$$

Where $\mu_r^{'}$=1 represents the mirror area relative permeability, μ_r=1.15 represents the relative permeability of the magnetic fluid. Under these circumstances, residual magnetic vector of primary magnet J_0 is 890 kA/m and that of mirror magnet J_{img} is 42.831 kA/m. This value is substituted into the simulation model as a parameter.

3.1.2 *Simulation and experiments.* As shown in figure 3(a), the second-order levitation force is converted to the repulsive force of the permanent magnet and the two mirror magnets in sealed container. However, in open container, since one end is a free surface, the levitation force is the repulsive force between the suspended permanent magnet and one mirror magnet, as shown in figure 3(b). In order to reduce the amount of calculation, levitation force of permanent magnet in open container is simulated. The result is further superimposed to obtain the levitation force of the permanent magnet in the sealed container.

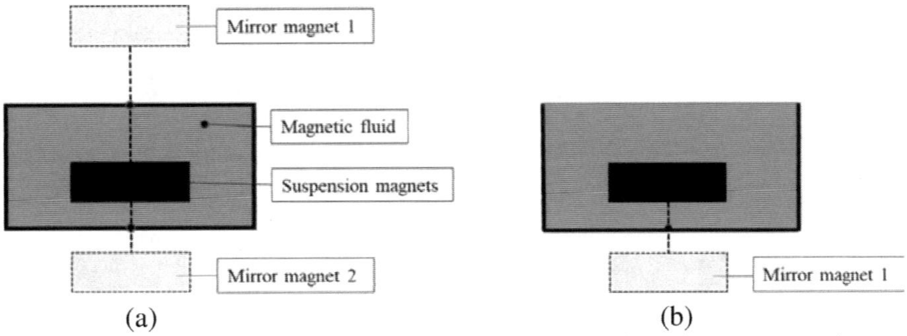

Figure 3. Levitation force diagram (a) sealed container (b) open container

Simulation is conducted based on finite element analysis software: Comsol Multiphysics. Figure 4 shows free split tetrahedral meshing of the simulation model. Since the container is bilaterally symmetrical, the forces in the x and y directions are balanced with each other. Only the force in the z direction is calculated. The distance is set between the two permanent magnets to 0~5mm at the interval of 0.1mm.

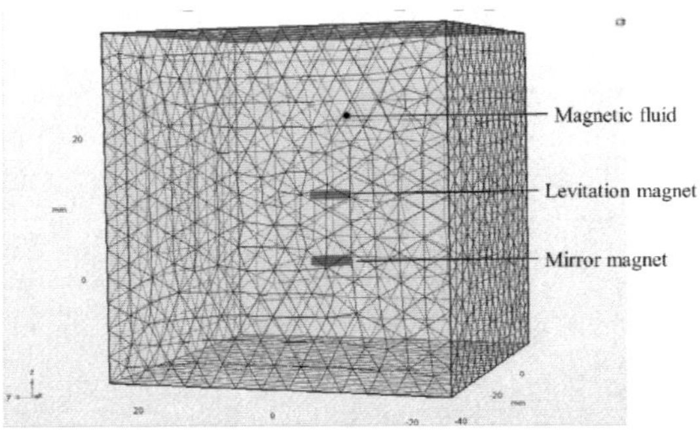

Figure 4. Free split tetrahedral meshing of the simulation model.

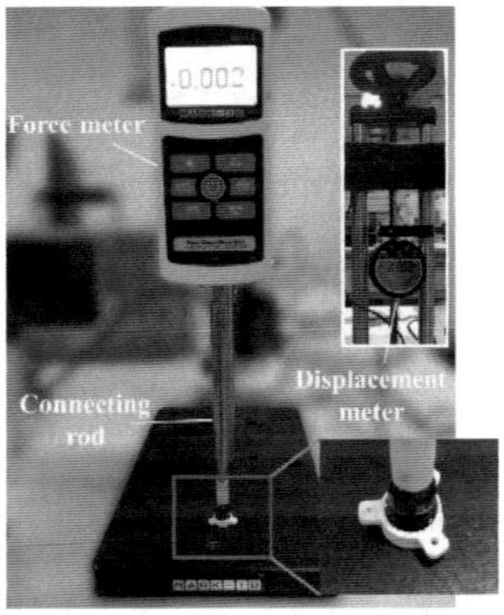

Figure 5. Bench diagram of levitation force experiments.

After the simulation results are obtained, the levitation force under the actual model is studied through experiments. As illustrated in figure 5, the magnet is attached to the end of the connecting rod which is connected to the push-pull force meter. Force meter is fixed to the displacement stage. During the experiment, the stage was first adjusted downward until the magnet was completely immersed in the magnetic fluid, and the push-pull force gauge was set to zero. Therefore, the influence of the Archimedes buoyancy is removed. The height of the stage is adjusted according to the displacement read through the displacement meter. As a result, the levitation force of magnet at different immersion heights can be obtained.

Since there is no analytical relationship between the levitation force and the levitation distance, the simulation and the experimental data are polynomial fitted. The fitting graph is shown in figure 6.

Figure 6 shows that the experimental and simulation data are basically in the same order of magnitude. As the levitation distance increases, the levitation force and the rate of its change gradually decreases. Since the levitation force is affected by many complicated factors including the magnetic field and actual parameters of the magnet. Simulation parameters are too difficult to be consistent with that in experiments because of the manufactured error. Many factors in the simulation cannot be fully considered, resulting in some errors between the simulation and experimental results. However, in order to judge the frequency expansion degree of the sensor after using magnetic fluid, this paper mainly focuses on the order of magnitude of levitation force research result. In this respect, the experimental and simulation results are in agreement. The results show that the levitation force of magnetic fluid is so small that it can be ignored in expansion of available frequency range. It will be used to offset the gravity of permanent magnet so as to eliminate the zero drift of the sensor.

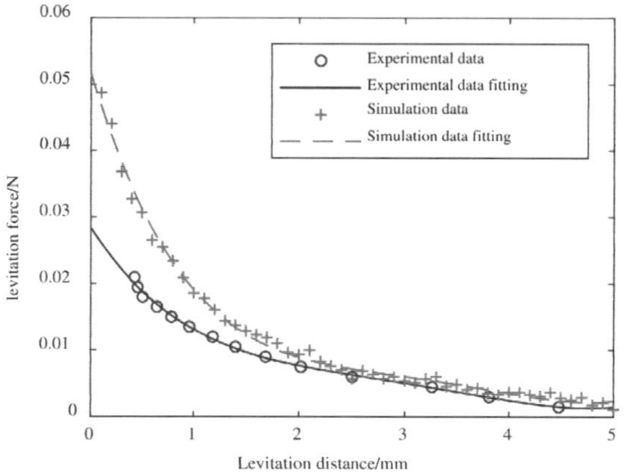

Figure 6. Fitting graph of experimental result

3.2. Damping effect

Damping effect means that damping ratio of vibration system using magnetic fluid as damping material can be influenced by magnetic field. Damping coefficient is a dynamic parameter. It is too difficult to measure it because magnetic fluid and permanent magnet are sealed in the inertial sensor. In this paper, vibration attenuation method is used to study the damping coefficient of magnetic fluid. Vibration attenuation curve is measured after tapping the sensor. Spring stiffness and the system damping ratio are calculated according to equations (7) (8).

$$\frac{1}{j}\ln\frac{A_i}{A_{i+j}} = \frac{2\pi\xi}{\sqrt{1-\xi^2}} \tag{7}$$

$$f = \frac{j}{T_{i+j} - T_i} \tag{8}$$

Where A_i and A_{i+j} represent two peak amplitude with j-th oscillation, T_i and T_{i+j} represent the corresponding time, ξ represents damping ratio.

Figure 7 shows the results of a single tap test on the inertial sensor without magnetic fluid. According to the attenuation curve and equation (7) (8), natural frequency and damping ration are obtained. The same method is used in the inertial sensor with magnetic fluid sealed inside. The result is shown in table 3.

In order to choose the best damping ratio for inertial sensor, experiments are repeated under sensors with different springs and magnetic fluid. Based on the experimental results, suitable spring and magnetic fluid are chosen to adjust the damping ratio to 0.56. This damping ratio can make available frequency range the widest.

Experimental results show that the distance between permanent magnet and the sensor casing has an effect on the damping effect of magnetic fluid. In addition to this, the magnetic field acted on the magnetic fluid is the key factor to affect the damping effect. Experiments introduced in this section guide the sensor structure design.

Table 3. The detailed information of components.

Condition	Natural frequency /Hz	Damping ratio
Without magnetic fluid	2029	0.0735
With magnetic fluid	1852	0.1729

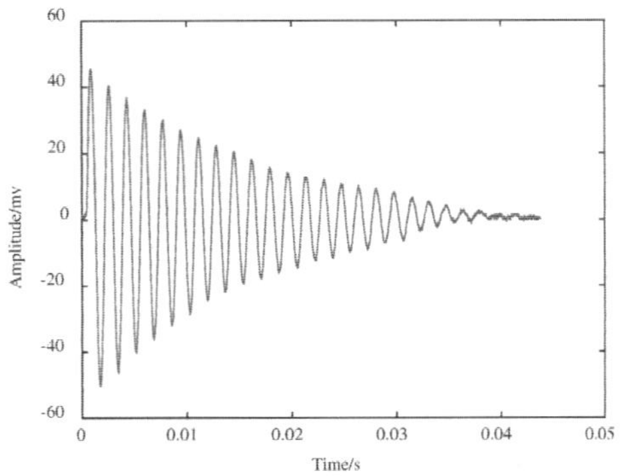

Figure 7. Vibration attenuation curve of the sensor without magnetic fluid

4. Verification experiments

Hall element is the measurement component of the inertial sensor. Through simulation and experiments, the position where magnetic field has best gradient is chosen to install the Hall element. It can make sensitivity of the inertial sensor best. This part will not be introduced in detail because it is not the focus of this paper.

Base on the results obtained in previous sections, parameters of sensor structure are determined. The designed sensor is manufactured and verification experiments are conducted on the vibration platform, which can provide 50g (gravity acceleration) at 30kHz. Figure 8 shows the manufactured sensor. The diameter of the sensor is smaller than 1cm. As shown in Figure 9, the standard acceleration sensor and the magnetic fluid sensor are mounted on the vibration platform.

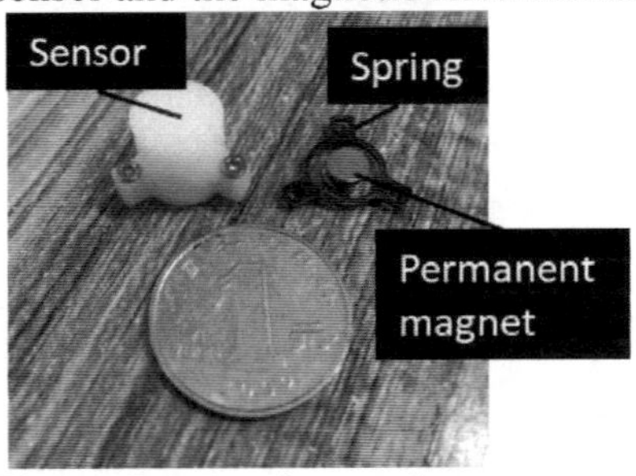

Figure 8. Manufactured sensor

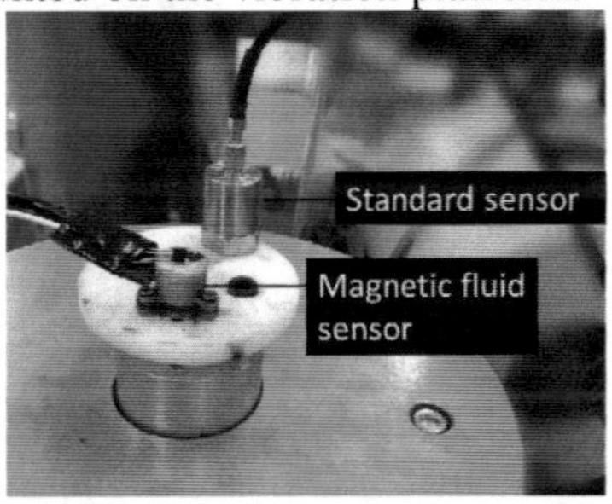

Figure 9. Verification experiments platform

The vibration platform is controlled to vibrate at different acceleration amplitudes and frequencies. At a frequency of 100 Hz, the acceleration amplitudes is controlled from $20\,\mathrm{m\cdot s^{-2}}$ ~$60\,\mathrm{m\cdot s^{-2}}$ with the interval of $5\,\mathrm{m\cdot s^{-2}}$. Figure 10 shows the sensor output. It indicates that the sensor designed in this paper has a good linearity in response to acceleration.

In order to verify the effect on the sensor brought by magnetic fluid, the frequency response experiments are also conducted on the vibration platform. The vibration platform is controlled to vibrate from 160Hz to 750Hz with the interval of 12.5Hz at the amplitude of $35\,\mathrm{m\cdot s^{-2}}$. According to second-order frequency response function, the experimental results are fitted. As illustrated in figure 11, the available frequency range of the sensor without magnetic fluid is 0~161.6Hz with the allowable error as 10%. When the sensor is added by magnetic fluid, the frequency range is 0~575.5Hz. The available frequency range is extended by 256%, which indicates that magnetic fluid has perfect properties to improve inertial sensors.

Moreover, the designed sensor is used to measure the rotor vibration in order to verify its practicality. The experimental platform is shown in figure 12. Standard sensor and magnetic sensor are installed at the same position of the rotor. The standard sensor used in this experiment is piezoelectric inertial sensor, which serves under 5kHz. The rotor is controlled to run at 24.2Hz. The results are compared in both time and frequency domain.

IOP Conf. Series: Materials Science and Engineering **625** (2019) 012022 doi:10.1088/1757-899X/625/1/012022

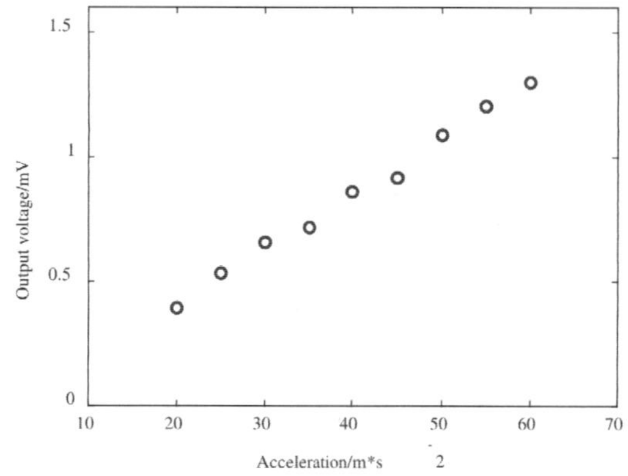

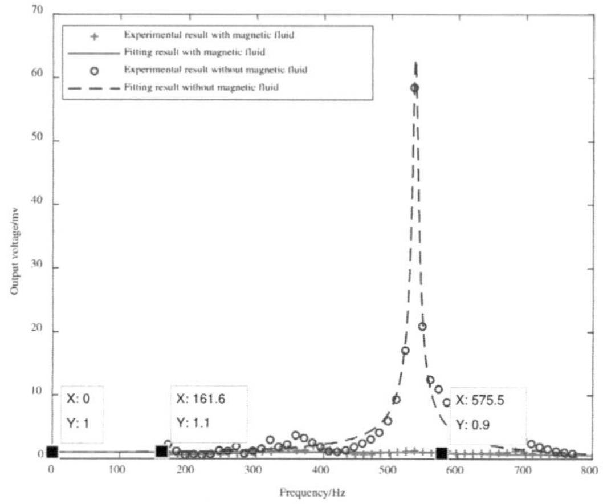

Figure 10. Sensor output in response to acceleration. **Figure 11.** Frequency response of the designed sensor

As illustrated in figure 13, information in time and frequency domain gained from two sensors are similar. It reflects that the magnetic sensor designed in this paper is capable of measuring the rotor vibration in reality. From figure 13(b), the rotary frequency of the rotor from magnetic fluid sensor is much clearer than that from standard sensor. It proves that magnetic fluid sensor has better performance than standard sensor in low frequency.

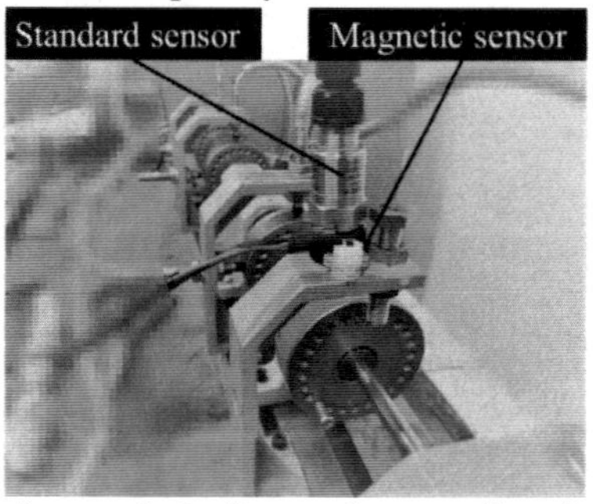

Figure 12. Experimental platform of rotor vibration measurement

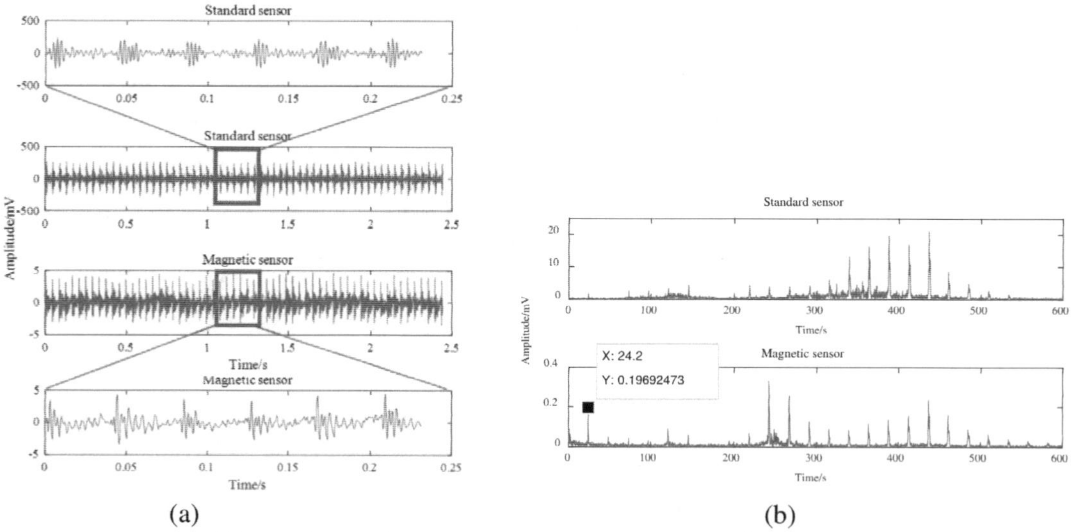

(a) (b)

Figure 13. Experimental result on rotor (a)time domain (b)frequency domain

5. Conclusions

A novel inertial sensor using magnetic fluid is designed in this paper. Simulation and experiments are conducted to study the characteristics of magnetic and the feasibility of using it in inertial sensor. The results show that magnetic fluid can be used to improve the available frequency range of inertial sensor. The innovations of this study mainly lie in: 1) Novel inertial sensor structure using magnetic fluid is designed to verify the feasibility to use magnetic fluid in inertial sensor. The simple structure keeps the sensor in low cost and good performance. 2) The levitation and damping effect of magnetic fluid are studied. Results show that magnetic fluid has good damping effect under the action of magnetic field. When the permanent magnet is small, the suspended force is so small that it can be ignored. 3) The innovative sensor structure and intelligent material magnetic fluid are combined. Experimental results show that the designed sensor has good performance in low frequency and magnetic fluid has a positive effect on extending the available frequency range.

This paper provides a strategy for studying the levitation and damping effect of magnetic fluid. It also gives a new idea for the research and improvement of acceleration sensors.

Acknowledgments

The authors gratefully acknowledge the financial support by the National Natural Science Foundation of China (Grant Nos.51275379 and 51421004).

References

[1] Zahn M 2001 Magnetic Fluid and Nanoparticle Applications to Nanotechnology. *J. Nanopart. Res.* **3** 73-8

[2] Zhang X, Xia X, Xiang Z, You Y and Li B 2018 An Online Active Balancing Method Using Magnetorheological Effect of Magnetic Fluid. *J. Vib. Acoust.-Trans. ASME* **141** 011008--9

[3] Barbour N and Schmidt G 1998 Inertial Sensor Technology Trends. *IEEE Sens. J.* **1** 332-9

[4] Qian L, Li D and Yu J 2016 The Theoretical and Experimental Study of a Novel Two-Dimensional Magnetic Fluid Accelerometer. *IEEE Sens. J.* **16** 8412-9

[5] Yao J, Huang C and Li D, 2016 Research on a Novel Ferrofluid Inertial Sensor with Levitating Nonmagnetic Rod. *IEEE Sens. J.* **16** 1130-5

[6] Piso M I 1999 Applications of Magnetic Fluids for Inertial Sensors *J. Magn. Magn. Mater.* **201** 380-4

[7] Qian L, Li D and Zhu T 2014 Use of Magnetic Fluid in Accelerometers *J. Sens.* **2014** 375623

[8] Yao J, Chen Y, Li Z, Zhang T and Li D A Novel Accelerometer Based on the First Kind of Ferrofluid Levitation Principle *Smart Mater. Struct.* **25** 095016

IOP Conf. Series: Materials Science and Engineering **625** (2019) 012023 doi:10.1088/1757-899X/625/1/012023

Synthesis of carbon nanoparticles using various electrolyte concentrations in submerged glow discharge plasma

M S Ranjit and M R M Julaihi

Swinburne University of Technology Sarawak, Jalan Simpang Tiga, 93350 Kuching Sarawak, Malaysia

Email: 100073374@students.swinburne.edu.my

Abstract. This research aims to study the effects of the manipulation of the concentration of electrolyte on the synthesis and size control of carbon nanoparticles. Submerged glow-discharge plasma (SGDP) method was used which is a type of electrolysis. This method was proven to be both cost effective and eco-friendly in the synthesis of nanoparticles. Carbon nanoparticles generally range from 1 to 100nm and its small size gives it a high surface area to low volume ratio which gives it an edge over the sheet and bulk metals. They vary in their mechanical properties with respect to their sizes which makes it ideal for the manufacturing industry. The ability to control carbon nanoparticle size will give manufacturers the ability to obtain the desired mechanical properties. This research brings the potential of being able to control the size of carbon nanoparticles with a simple and cost-effective method by changing the concentration of electrolyte and plasma discharge voltage. The manufacturing industry in the future can benefit from this research as they will be able to manufacture materials with better mechanical properties at a cheaper cost.

1. Introduction

Synthesis and size control of nanoparticles require considerable amount of research. The main goal is being able to control the size of these nanoparticles depending on their applications. The method used here is the submerged glow- discharge plasma (SGDP) method. SGDP was proven to be simple to set up, does not require hazardous materials and cost effective [1]. The parameters that can be controlled in SGDP method are categorized into three categories (2):

1) Manipulating the concentration of the electrolyte.
2) Manipulating the depth of the cathode.
3) Manipulating the type of electrolyte used.

In this paper, a variety of electrolyte concentrations were used with the prospects of controlling the size of the synthesized carbon nanoparticles. Each of the carbon nanoparticles formed from the SGDP method is expected to show different mechanical properties based on the variation of their size with respect to the concentration of the electrolyte. Controlling the size of carbon nanoparticles can bring about controlling the properties of the nanoparticles produced. Being able to manufacture carbon nanoparticles by the numbers with a low starting cost shows the potential of this method being implemented in the future. Furthermore, SGDP is eco-friendly as no unwanted emission occurs except for water vapor when carrying out this process, which helps in the reduction of environmental pollution [2-3]. The manipulation of the depth of the cathode and the type of electrolyte used on the synthesis and size control of the carbon nanoparticles will be under the future research section.

2. Submerged glow-discharge plasma method (SGDP)

The SGDP method is a form of electrolysis of water which comprises of four main stages; the electrolysis stage, vaporization stage, transition stage and plasma stage. It revolves around the collision of electrons and neutral atoms on the cathode surface which causes a spark which excites, ionize and dissociates with the emission of light "glow phenomenon" [4-5]. The schematic diagram summarizing the formation of nanoparticles by SGDP is shown in Figure 1.

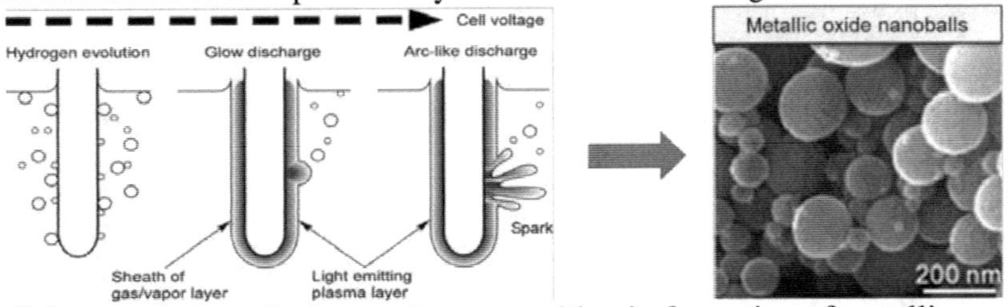

Figure 1. Submerged glow discharge plasma resulting in formation of metallic nanoparticles

2.1 Carbon nanoparticles

Carbon nanoparticles are particles which range from 1 – 100 nm. Carbon nanoparticles are made up of graphite which consists of intermediate crystalline structure. The range setting was done to differentiate nanoparticles and cluster particles.

2.2 Size control of carbon nanoparticles

The main focus for controlling the size of the carbon nanoparticles was to vary the concentration of the electrolyte. Different concentrations of electrolyte were expected to result in the change of carbon nanoparticles size and morphology. Using potassium carbonate (K_2CO_3) as the electrolyte, increasing the electrolyte concentration would increase the electrical current value as well. The high concentration of electrolytes indicates that there were higher numbers of free-flowing electrons in the solution. This in turn would increase the conductivity of the solution, subsequently increasing the current value at low voltages. This increases the number of substance in the solution [2]. In terms of size, it is predicted that a lower glow discharge voltage would decrease the size of the carbon nanoparticles produced [2,6].

The carbon nanoparticles formation on the cathode will vary in their sizes depending on the concentration of the electrolyte. A higher concentration electrolyte will provide more free-flowing electrons (e^-), thus the probability of achieving glow-discharge voltage would be higher. At the same time more electron bombardment can occur on the cathode surface. Therefore, achieving the glow-discharge voltage or plasma initiation at a quicker rate will cause the sizes of the carbon nanoparticles to decrease. The electrolyte used here was potassium carbonate (K_2CO_3) which is a strong electrolyte. As a strong electrolyte, it has better electrical conductivity compared to weak electrolytes [7].

3. Methodology

3.1 Experimental setup

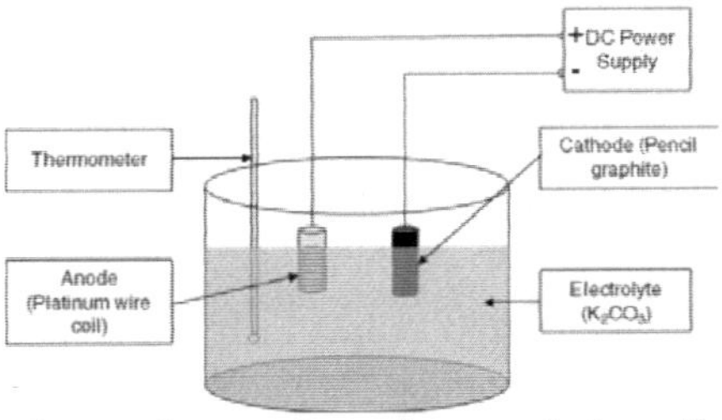

Figure 2. Experimental setup for submerged glow discharge plasma

IOP Conf. Series: Materials Science and Engineering **625** (2019) 012023 doi:10.1088/1757-899X/625/1/012023

3.2 Experimental procedure

The apparatus was set as shown in Figure 3. An 800W DC power supply (Kikusui PWR800L) was used to generate a steady electrical current supply to the setup. A 100mm long graphite rod obtained from a mechanical pencil refill was used as the cathode and a 500mm coiled platinum wire was used as the anode. The concentration of the K_2CO_3 electrolyte was varied to 0.1M, 0.5M and 1.0M with different voltages applied. Scanning electron microscopy (SEM) was used to obtain images of the nanoparticles, while energy dispersive spectroscopy (EDS) was used to determine the elemental composition of the nanoparticles.

4. Results and discussion

The voltage and current measurements during plasma generation for each of the electrolyte is shown in Table 1.

Table 1. The plasma generation conditions at 0.1M, 0.5M and 1.0M K_2CO_3

Concentration (M)	Sustained voltage (V)	Current (A)
0.1	100	1.5
0.5	75	1.5
1.0	15	4.5

The mass of the carbon nanoparticles are shown in Table 2. The mass of the nanoparticles were calculated as the difference in mass of the cathode before and after plasma discharge, assuming that any loss of the cathode mass was due to it converted into nanoparticles

Table 2. Mass of graphite rod before and after plasma discharge

Concentration, K_2CO_3 (M)	Initial mass of graphite, m_1 (g)	Mass of graphite after plasma process, m_2 (g)	Nanoparticle mass, $m_1 - m_2$ (g)
0.1	125	120	5
0.5	125	114	11
1	125	110	15

As the electrolyte concentration increased, the mass difference of the graphite rod before and after plasma discharge became larger. This shows that the graphite rod was converted into carbon nanoparticles when the SGDP method was carried out. Higher concentrations lead to more bombardments of electron on the wall of the cathode which justified the decrease in the mass of the graphite rod.

Table 3. The atomic percentage of elements of the synthesized nanoparticles

Concentration (M)	C (%)	O (%)
0.1	85.39	14.61
0.5	77.97	22.03
1.0	89.45	10.55

Table 3 shows the EDS results conducted on the particles that were obtained after SGDP process. It is clear that the elemental compositions for all of the nanoparticles show that they consisted of mainly carbon. The presence of oxygen showed the possibility that some of the nanoparticles were oxidized.

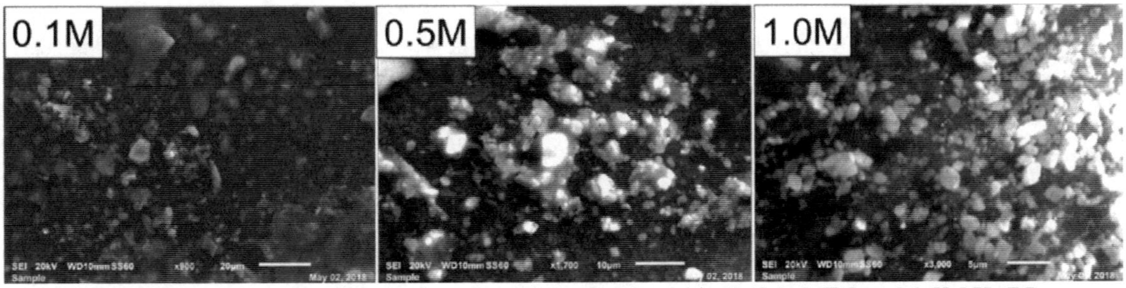

Figure 3. SEM images of carbon nanoparticles produced in 0.1M, 0.5M and 1.0M K_2CO_3

Figure 3 shows that the size of carbon nanoparticles produced varied with different electrolyte concentration. In 0.1M and 0.5M, the particles were quite large and were produced in small amounts compared to 1.0M electrolyte. Meanwhile, in at 1.0M concentration, there were numerous smaller nanoparticles compared to 0.1M and 0.5M concentrations. Here, the higher the electrolyte concentration and thus the lower the glow discharge voltage, the carbon nanoparticles were more uniform in shape and smaller in size

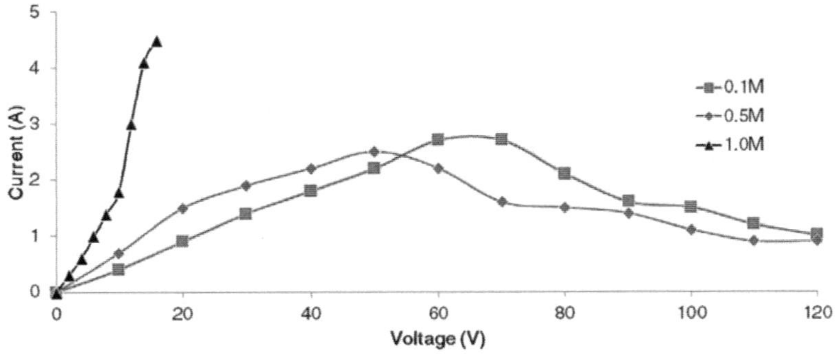

Figure 4. Voltage-current relationship during plasma generation on graphite cathode

Figure 4 shows the voltage-current relationship during plasma generation for each of the electrolyte concentrations. For an electrolyte concentration of 1.0M K_2CO_3 the glow discharge voltage was achieved at 15V producing a current of 4.5A while the 0.1M K_2CO_3 electrolyte concentration which achieved its glow discharge voltage at 100V producing a current of 1.5A. According to Figure 4, the higher the concentration of the electrolyte, the higher the current flow in the aqueous solution. The glow-discharge voltage in turn was achieved at lower voltage.

The concentration of the electrolyte determines the number of free flowing electrons within the solution. More free flowing electrons meant that the glow-discharge voltage would be attained at a lower voltage. Having more free flowing electrons enabled more secondary electron bombardment on the walls of the cathode. This kept the plasma self-sustaining as well as producing large amounts of small sized carbon nanoparticles.

The main effect on carbon nanoparticles sizes formed was due to the electrolyte concentration influencing the glow discharge initiation voltage. This in turn affected the plasma sustaining voltage. The lower the glow discharge sustaining voltage the smaller the size of the carbon nanoparticles produced. The concentration of the electrolyte determines the number of electrons to bombard the walls of the cathode. More electrons being able to collide with the wall of the cathode to produce the self-sustaining plasma the lower the glow discharge initiation and sustaining voltage which prompts the release of smaller sized carbon nanoparticles.

5. Future research

Manipulating the depth of the cathode rod and the type of electrolyte used are two further parameters that are important in identifying the most optimal way of controlling the size of the carbon nanoparticles produced.

Manipulating the depth of the cathode rod while maintaining the concentration of the electrolyte in the solution is to test how the glow-discharge voltage will vary within this condition. Ideally, the increase in the surface area exposed to the electrolyte is expected to increase the glow-discharge voltage. This correlates to the production of carbon nanoparticles which depends on the plasma being able to keep its self-sustaining state. Having a larger surface area meant that the secondary electrons bombarding the walls of the cathode would be more scattered across the surface of the cathode, this is expected to cause the glow-discharge voltage to increase thus increasing the size of the carbon nanoparticles formed. A physical experimentation is to be done in order to verify this assumption.

Manipulating the electrolyte used will also indirectly influence the glow-discharge of voltage in the solution. In theory, using a stronger electrolyte will yield more accurate results as the electrolyte will completely ionize in water. Stronger electrolyte will have better electrical conductivity, thus is expected to decrease the glow-discharge voltage. This was confirmed through an experimentation carried out by Al Anbouri [4]. Despite this, not a lot is known on how to optimally control the size of the carbon nanoparticles through using the different types of electrolytes.

6. Conclusion

Submerged glow discharge plasma method was used to produce carbon nanoparticles. The carbon nanoparticles were found to be more uniformly shaped and smaller at 1.0M compared to 0.1M and 0.5M electrolyte concentrations. The increase in the concentration of the electrolyte affects the glow discharge voltage which controlled the size of the carbon nanoparticles formed. This breakthrough will be very beneficial to the manufacturing industry, as they will be able to control the size of the carbon nanoparticles through manipulation of the size of the electrolyte concentration and plasma discharge voltage. Furthermore, the SGDP method is also a simple and cost effective method of carrying out the production carbon nanoparticles.

References

[1] Ishigami M, Cumings J, Zettl A, Chen S 2000 Chem Phys Lett. 319 457
[2] Al Anbouri MHS, Yii BWM, Yek PNY, Julaihi MRM 2018 IOP Conf. Ser.: Mater. Sci. Eng. 429 01208
[3] Yek PNY, Julaihi MRM, Osman MS, Tiong TC, Lee WH, Lee CL 2018 E3S Web of Conferences (CENVIRON 2017) 34 01028
[4] Julaihi MRM, Yatsu S, Watanabe S 2014 J Exp Nanosci 10 965
[5] Kim SW, Lee S, Saqib ANS, Lee YH, Jung MH 2017 Curr Appl Phys. 17 181
[6] Pootawang P, Saito N, Lee SY 2013 Nanotech 24 055604
[7] Saito G, Wan Mohd Azman WOS, Nakasugi Y, Akiyama T 2014 Adv Pow Tech 25 1038

Synthesis of calcium phosphate composite organogels by using castor oil and sorbitan monopalmitate based for dentine occlusion material

J Waiyawat[1], K Akkarachaneeyakorn[1], P Changpongpan[1], M Kongsema[2], C Teanchai[3] and O Chienthavorn[1]

[1] Department of Chemistry, Faculty of Science, Kasetsart University, Bangkok, Thailand 10900
[2] Department of Zoology, Faculty of Science, Kasetsart University, Bangkok, Thailand 10900
[3] Faculty of Dentistry, Mahidol University, Bangkok, Thailand 10400

Email: jutharat.wai@ku.th

Abstract. Calcium phosphate composite organogel was synthesized from a mixture of sorbitan monopalmitate as emulsifier and castor oil as oil phase at 60 °C to establish emulsion system with calcium chloride and di-sodium hydrogen phosphate aqueous solution as water phase. The heated mixture was cooled at room temperature to form organogel. The prepared organogels were investigated by thermogravimetric analysis (TGA), scanning electron microscope (SEM), transmission electron microscope (TEM) and cytotoxicity test. The prepared organogels of 4:3 and 1:1 in oil/water content ratio were white and gritty in nature, which resulted in calcium phosphate nanoparticles with spherical shape of 1.3 to 1.9 μm and 1.2 to 2.3 μm, respectively. They were able to successfully occlude into dentine tubules within 1 day. In addition, the prepared organogels were found to have acceptable toxicity level for daily human use with high concentration up to 1600 μg/ml. Hence, these organogels showed the promising results to be developed as an effective dentine occlusion material in relieving of dentin hypersensitivity.

1. Introduction

Tooth sensitivity also known as dentin hypersensitivity is a prevalent problem of adult. According to previous researches, one in seven patients (or 8 to 57%) presented in a dental clinic to treat the pain from sensitive teeth [1]. Hypersensitivity symptom is typically transient sharp pain caused by thermal, chemical, evaporative, tactile, osmotic, or mechanical stimuli on exposed dentinal tubules [2]. Exposure of the dentine to cause the hypersensitivity occurs from several factors, for example a removal of tooth enamel occurred by the use of a hard toothbrush, gum recession, tooth erosion resulted from consuming acidic foods, an effect of gastroesophageal reflux disease which then leaves the dentine tubule to expose, or their combinations [3]. To reduce tooth sensitivity, in this research a desensitizing product will be developed on the basis of dentine tubule occlusion by using bioactive organogel. Since hydroxyapatite (HAp,$Ca_{10}(PO_4)_6(OH)_2$) is not only the major component of bone, dental enamel and dentine [4], but also the most stable crystalline form of calcium phosphate under human body conditions as pH, temperature and composition of body fluids, the prepared organogel was then embedded with hydroxyapatite [5,6].

Castor oil is an unsaturated fatty acid, containing ricinoleic acid (12-hydroxy-cis-octadec-9-enoic acid) that can be extracted from seed of *Ricinus communis*. It is popular in various industries, such as pharmaceutical products, cosmetics and synthetic detergents, owing to its anti-inflammatory, anti-oxidant and anti-microbial activities [7]. Sorbitan monopalmitate (SMP) is a non-ionic surfactant, containing palmitic acid and sorbitol in the chemical structure. As SMP can facilitate the formation of emulsion, it has been widely employed in various pharmaceutical and cosmetic products [8]. In this study, the synthesis of calcium phosphate composite organogel was performed by using the mixture of sorbitan monopalmitate and castor oil based with modified process from previous report [8]. Different ratios of oil/water were varied to study the effect of formation of organogels and structure of synthesized calcium phosphate nanoparticles. The prepared calcium phosphate composite organogels were characterized and investigated for their toxicities and abilities to occlude the dentine tubule on human dentin slices.

2. Experimental section

2.1 Materials

In synthesis of calcium phosphate composite organogel, sorbitan monopalmitate (SMP) (MW 402.57 g/mol) and castor oil were obtained from Sigma-Aldrich (Saint Louis, USA). Calcium chloride monohydrate ($CaCl_2 \cdot H_2O$) (MW 147.02 g/mol) and di-sodium hydrogen phosphate (MW 141.96 g/mol) were bought from Merck KGaA (Darmstadt, Germany). All reagents were used without further purification.

2.2 Preparation of calcium phosphate composite organogels

Organogels were prepared according to previous report [8] with a slight modification. Initially, 800 µl of castor oil was heated at 60 °C and added 0.23 g of sorbitan monopalmitate with stirring at 500 rpm for dispersion as the hot solution was transparent. A 100 µl of calcium chloride solution (7.4×10^{-4} moles) was dripped into the solution and stirred for 30 min at 700 rpm. An aqueous solution of di-sodium hydrogen phosphate solution (7.2×10^{-4} moles in 500 µl of water) was then added dropwise with stirring for 1 h. The hot solution was cooled down to ambient temperature to form gel 1 with [Ca]:[P] molar ratio of 1:1. For gel 2 and 3 the volume of water was subsequently varied from 500 to 700 and 800 µl however the [Ca]:[P] molar ratio of gels are maintained the same.

2.3 Characterization

The calcium phosphate composite organogels were heated from 25 to 800 °C at a rate of 10 °C/min to determine the inorganic content by TGA. TEM investigation was also performed by using a Hitachi HT7700 transmission electron microscope (Hitachi High Technologies, Hitachinaka-shi, Japan). For the morphology and sizes studies, organogels were dispersed in ethanol and then sonicated for 15 min. A 2 µl of the dispersion liquid were dropped onto carbon - coated copper grids and left in a desiccator overnight. The analysis was performed at an accelerating voltage of 120 kV. The surface morphology of calcium phosphate nanoparticles in organogels was investigated by using a FEI Quanta 450 scanning electron microscope (Field Electron and Ion Company (FEI), Hillsboro, Oregon, USA) with 20 kV accelerating voltage.

2.4 Cytotoxicity test

Organogels were tested for cytotoxicity on human dermal fibroblast cell line. Briefly, cells were treated with various concentrations of the organogel, which was previously sterilized by UV exposure, followed by an incubation at 37 °C in a 95% humidified incubator with 5% CO_2 for 24 h. To investigate the cytotoxicity in cell, the sulforhodamine B (SRB) test was conducted. SRB was dissolved from living cells using 10 mM Tris buffer to observe the optical density at 492 nm using a Biotek PowerWave XS micro-plate reader (Biotek, Winooski, Vermont, USA).

2.5 Dentinal tubule occlusion

Dentine slices were prepared for tubule exposure. Each organogel was applied onto the surface of dentine specimen, then left for 1 day under moist condition. After that, organogel was washed away with water before sonication for 15 min. All of dentine specimens were kept in a desiccator before morphological study under SEM with using 15 kV accelerating voltage.

3. Results and discussions

The synthesis of calcium phosphate composite organogels were performed with the mixture of SMP and castor oil at 60 °C to form gel. Three ratios of castor oil/water were examined to obtain different texture of gel 1, 2, and 3. All prepared gels were white and turbid. The gel 1 and 2 were non-homogenous and gritty in nature, while the gel 3 was more homogeneous, as it had the highest water content. The TGA profile for gel 1, 2 and 3 shows weight losses of water of 29.3, 31.2, and 46.6%, respectively, at below 100 °C (Figure 1). The temperature was much lower than that for similar material of 150 °C in the report of Wu *et.al.* [9]. The sharp change in the range of 280 to 420 °C was attributed to the decomposition of organic group, which typically occurs between 300 to 800 °C [9]. After the decomposition of gel 1, 2 and 3 at 800 °C the final weights were 6.6, 7.8 and 8.7%, respectively, which indicated inorganic products of calcium phosphate composites. The SRB assay was utilized for cytotoxicity test, according to the ISO 10993 parts 5 and 12, the prepared organogel shows the minor toxicity (toxicity grade 1, 75 to 99% of relative growth rate [10, 11]) at a high concentration of the organogel up to 1600 μg/ml (Figure 2).

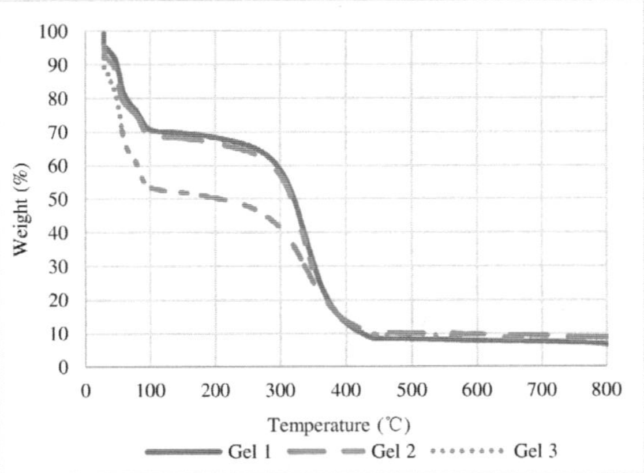

Figure 1. TGA curves of calcium phosphate composite organogels.

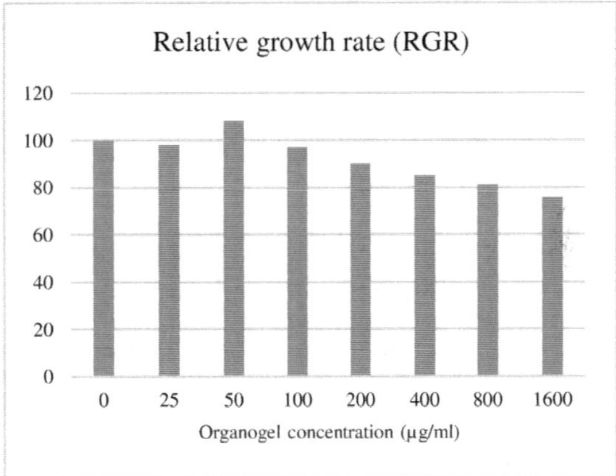

Figure 2. Cytotoxicity test of calcium phosphate composite organogels.

The morphologies of calcium phosphate nanoparticles were investigated by TEM as shown in Figure 3. The nanoparticles presented spherical-like shape, which was very similar to the emulsion system investigated in the previous study [12]. The gel 1 contained particles of 1.3 to 1.9 μm diameter partly adhered with very small granules (Figure 3a). The gel 2 exhibited the nanoparticles with a dimension range of 1.2 to 2.3 μm (Figure 3b), while the gel 3 gave particles with an increased size of 2.1 to 3.6 μm (Figure 3c). Similarly, the morphology of prepared organogels confirmed by SEM images shows spherical calcium phosphate nanoparticles dispersing within the gel matrix (Figure 4).

To study ability as a bioactive filler material for dentine tubule occlusion the prepared calcium phosphate composite organogels were also examined on human dentin slice. According to SEM studies, gel 1 and 2 showed the efficiency to occlude the exposed dentinal tubules within 1 day by calcium phosphate nanoparticles (Figure 5a and 5b). However, the tooth slice of gel 1 treatment exhibited partial tubule exposure from the incomplete occlusion. Owing to the size of dentine tubules with approximately of 1.7 - 2.4 μm (Figure 5c), the sizes of calcium phosphate nanoparticles in gel 3 were larger than the sizes of tubules. Therefore, the treatment of gel 3 was not be able to occlude tubules within 1 day.

IOP Conf. Series: Materials Science and Engineering **625** (2019) 012024 doi:10.1088/1757-899X/625/1/012024

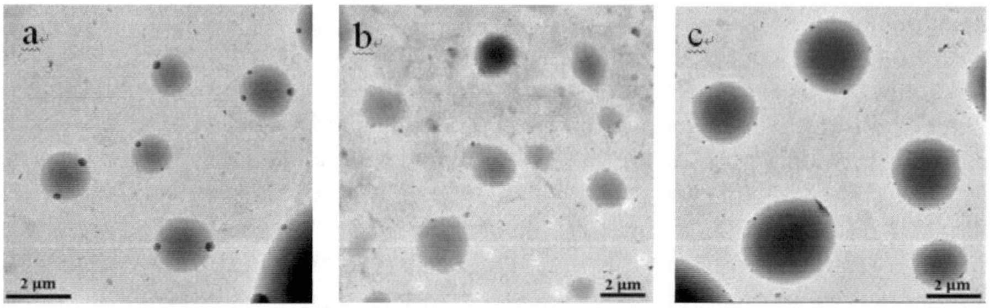

Figure 3. TEM images of calcium phosphate structures (a) gel 1, (b) gel 2 and (c) gel 3.

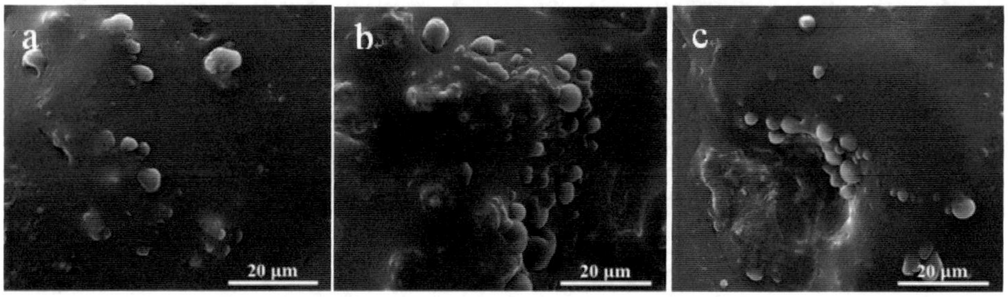

Figure 4. SEM images of calcium phosphate structures (a) gel 1, (b) gel 2 and (c) gel 3.

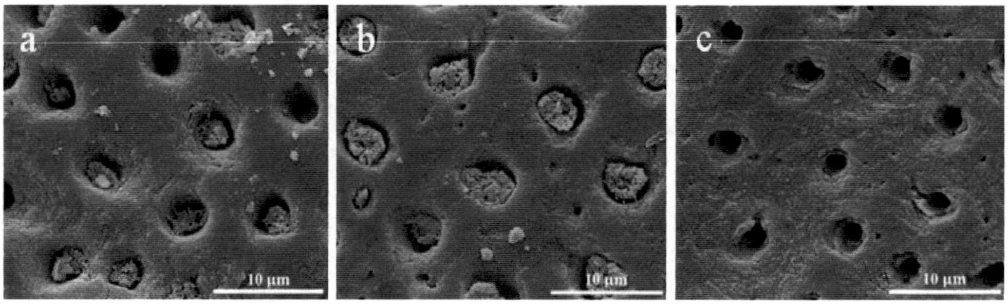

Figure 5. SEM images of dentine surface treated with (a) gel 1, (b) gel 2 and (c) gel 3 for 1 day, showing tubular occlusion of dentine by calcium phosphate.

4. Conclusion

In this work, the synthesis of calcium phosphate composite organogels for dentine tubule occlusion were successfully prepared via emulsion method using the mixture of SMP-castor oil base. The texture of organogels were either gritty or smooth white in nature depending on composition of the gel formulation. The calcium phosphate nanoparticles dispersed in the gel 1 and 2 matrix were spherical with the dimension range of 1.3 to 1.9 and 1.2 to 2.3 μm respectively. Both gels were able to plug into dentine tubules within 1 day. Moreover, the high concentration up to 1600 μg/ml of organogel was acceptable toxicity level for human. The results confirmed that the prepared calcium phosphate composite organogels can be effectively used as a bioactive filler material in the treatment of dentine hypersensitivity. However, the work should be developed continually for treatment in vitro with artificial saliva and acid test to examine the wear resistance and stability of materials. The transverse sectioning of filled dentine tubules should be perform to observe the penetration depth of deposited calcium phosphate. The different mole ratio of calcium/phosphate should be varied to improve particles size and structure prior to being used as the effective bioactive filler material.

Acknowledgements

This work is supported in part by grants from the Faculty of Science, Kasetsart University (Thailand), Kasetsart University Research and Development Institute (KURDI), Chemistry Research Credit (CRC) from the department of chemistry and National Science and Technology Development Agency

(NSTDA). The Center of Excellence for Innovation in Chemistry (PERCH-CIC), Office of the Higher Education Commission, Ministry of Education is acknowledged for the scholarship for J. Waiyawat.

References

[1] Fischer C, Fischer RG, Wennberg A 1992 Brazil. *J. Dentist.* **20(5)** 272-276.
[2] Shiau HJ. 2012 *J. Evid. Base. Dent. Pract.* **12(3)** 220-228.
[3] Won K, Ji, Park JC. 2017 *J. Oral Biosci.* **59(4)** 211-217.
[4] Bose S, Saha SK. 2003 *Chem. Mater.* ***15*(23)** 4464-4469.
[5] Karimzadeh A, Ayatollahi MR, Shirazi HA. 2014 *Int. J. Mater. Mech. Manuf.* **2(1)** 67-72.
[6] Das BB, Mitra A. 2014 *Int. J. Mater. Mech. Manuf.* **2(1)** 41-46.
[7] Kulkarni MG, Sawant SB 2003 *Eur. J. Lipid Sci. Technol.* **105(5)** 214-8.
[8] Singh VK, Pal K, Pradhan DK, Pramanik K 2013 *J. appl. Poly. Sci.* **130(3)** 1503-15.
[9] Wu Z, Xiang H, Kim T, Chun MS, Lee K 2006 *J. Colloid Interface Sci.* **304(1)** 119-24.
[10] Cao B, Zheng Y, Xi T, Zhang C, Song W, Burugapalli K, Yang H, Ma Y 2012 *Biomed. Microdevices* **14(4)** 709-720.
[11] International Organization for Standardization ISO10993-5 2009 Arlington *Biological evaluation of medical devices* VA: ANSI/AAMI.
[12] Bohner M, Tadier S, van Garderen N, de Gasparo A, Döbelin N, Baroud G 2013 *Biomatter* **3(2)** e25103.

IOP Conf. Series: Materials Science and Engineering **625** (2019) 012025 doi:10.1088/1757-899X/625/1/012025

Application of kepok banana peel activated carbon prepared by conventional and microwave heating for malachite green adsorption

W Astuti[1*], Megawati[1], MA Mahardhika[1], DA Putri[1], M Rohman[1], MF Sihab[1], T Sulistyaningsih[2], M Hidayah[1], L Fitriana[1] and EF Irchamsyah[1]

[1] Department of Chemical Engineering, Universitas Negeri Semarang, Semarang, Indonesia
[2] Department of Chemistry, Universitas Negeri Semarang, Semarang, Indonesia

*Email: widi_astuti@mail.unnes.ac.id

Abstract. Kepok banana peel as an agricultural waste has been used as a precursor in the production of activated carbon with NaOH activation. In this sense, the effect of heating technique (e.g. conventional and microwave heating) to the character and adsorption performance of the activated carbon was investigated. The obtained activated carbon was characterized by pore structural analysis, Fourier transform infrared (FTIR) spectroscopy and scanning electron microscope (SEM), then it used to remove malachite green from aqueous solution. It was found that microwave assisted NaOH activated carbon possessed high surface area and uniform pore structure. The surface area of activated carbons produced by conventional and microwave heating were 10.95 and 187.66 m^2/g, respectively. The optimum malachite green adsorption occurred at pH 10 and the equilibrium reached after 180 min. The adsorption behavior was well described by the Freundlich isotherm model.

1. Introduction

Wastewater released from textile, paint and printing industries generally contains a large number of dyestuff. If it enters into the water body, it can inhibit photosynthesis process of plants in the water [1]. The dyes have a complex aromatic structure that is stable to biodegradation and photodegradation. In addition, several dyes are toxic, carcinogenic and mutagenic. Considering the danger caused by the dyes, it must be removed from wastewater before discharged into the stream. Several methods such as electrochemical treatment [2], coagulation [3] and adsorption [4] have been used to remove dyes in aqueous solution. Among them, adsorption using activated carbon is the promising method due to its simplicity and high performance [5]. Recently, the use of lignocellulose-based activated carbon is preferred due to low cost and its abundant availability [6-8]. In this study, kepok banana (Musa paradisiaca) peel waste was used as a precursor in the preparation of activated carbon due to it has a high content of hemicellulose (74.9%), lignin (7.9%) and cellulose (7.5%).

In general, physical or chemical activation methods are used in the synthesis of activated carbon. In contrast to the physical activation where the resulting char is activated in the presence of CO_2 or steam [9], in chemical activation, char is impregnated with activating agents such as $ZnCl_2$ [10] and H_3PO_4 [11]. In this sense, conventional heating using a furnace is usually adopted. Recently, microwave technique becomes a promising alternative for conventional heating due to it has several advantages including uniform temperature distribution, shorter time and lower energy [12]. To the best of our knowledge, no literature have been reported on the preparation of kepok banana peel activated carbon by microwave-assisted NaOH activation. Therefore, the objectives of this study were to (1) prepare and characterize activated carbon from kepok banana peel with NaOH activation using conventional and microwave heating (2) investigate adsorption characteristics of malachite green onto the activated carbon and (3) determine isotherm model for malachite green-activated carbon system.

2. Materials and methods

2.1. Preparation of activated carbon

Kepok banana peel used as a precursor in this study was obtained from a traditional market in Semarang, Indonesia. The banana peel was washed and heated at 110°C in an oven to remove moisture content. The dried sample was then blended to a powder. Furthermore, char was produced through a carbonization process by heating the sample in a furnace at a temperature of 500°C for 2 hours. The obtained char was mixed with NaOH and 10 mL of distilled water in a vertical reactor and stirred for 2 hours at room condition. The ratio of NaOH : char was 3:1. Afterward, the sample dried at 130°C for 4 h. There are two heating techniques. First, using a furnace at a temperature of 700°C for 1 hour. Second, using a microwave oven with a frequency of 2.45 GHz at 600 W for 7 min, under a stream of N_2 with a flow rate of 0.25 L/min. The obtained activated carbon was washed with 0.1 M HCl and distilled water until the filtrate pH was around 7. Furthermore, it was dried at 110°C for one hour.

A scanning electron microscope ((Phenom Pro X Desktop, UK) was used to analyze the surface morphology of char, conventional heated activated carbon (CAC) and microwave heated activated carbon (MAC) while the surface area was characterized through adsorption of N_2 at 77K using gas sorption apparatus (Quantachrome, USA) and calculated by Brunauer-Emmet-Teller (BET) method. The surface organic structure of the adsorbents was characterized by Fourier transform infrared (FTIR) spectroscopy (Shimadzu, Japan) using KBr pellet method at wave number 400-4000 cm^{-1}.

2.2. Adsorption studies

Malachite green (chemical grade reagent) used as adsorbate was obtained from U.D. Indrasari Semarang and used without further purification. The sample solution was prepared by dissolving the malachite green powder with distilled water to the required concentration.

Batch adsorption experiments were conducted in six Erlenmeyer (100 mL size) containing 0.1 g of adsorbent and 50 mL of malachite green solution (100 mg/L). The adsorption was carried out at various pH (2, 4, 6, 8, 10 and 12) with added 0.1 N HCl or 0.1 N NaOH. The suspensions were shaken using a shaker apparatus at 120 rpm for 180 min. All suspended solids were filtered using Whatman paper and the filtrates were then analyzed for residual malachite green using a UV-Visible spectrophotometer at λ_{max} 618 nm. The effect of contact time toward the quantity of malachite green adsorbed was studied at various contact time (10, 20, 30, 40, 50, 60, 120, and 180 min) at solution pH of 10 and malachite green concentration of 100 mg/L. Adsorption isotherm was studied through various malachite green concentration (5, 10, 20, 30, 40, 50, 100, 200, 300, 400, and 500 mg/L) at solution pH of 10 for 180 min. The malachite green uptake at equilibrium per unit mass adsorbent (mg/g) was calculated by Eq. (1) as follows [13]:

$$q_e = \frac{(C_0 - C_e)\,V}{M} \tag{1}$$

where C_o and C_e (mg/L) are concentrations of malachite green in the liquid-phase at initial and equilibrium, respectively, M (g) is the mass of adsorbent used and V (L) is the volume of solution.

3. Results and discussion

3.1. Characterization of char, CAC and MAC

Figure 1 describes SEM micrographs of char, CAC, and MAC. It can be observed that the char (Figure 1a) has a dense and constricted surface due to the pores were blocked by tarry substances. While, after the activation process with NaOH, the porosity of the activated carbon increased (Figure 1(b) and 1(c)). In this sense, pore development through three stages including the opening of the blocked pores, widening of the existing pores and formation of new pores. In conventional heating, material (i.e. char) was heated by conduction and convection mechanism from a heat source located outside material bed. It led to a temperature gradient in the bed until a steady state condition was reached. As a result, the pore size was very diverse (Figure 1(b)). In contrast to the conventional heating, in the microwave heating, material received heat through dipole rotation and ionic conduction mechanism. As a result, MAC surface (Figure 1(c)) had a highly porous structure with more uniform pore size.

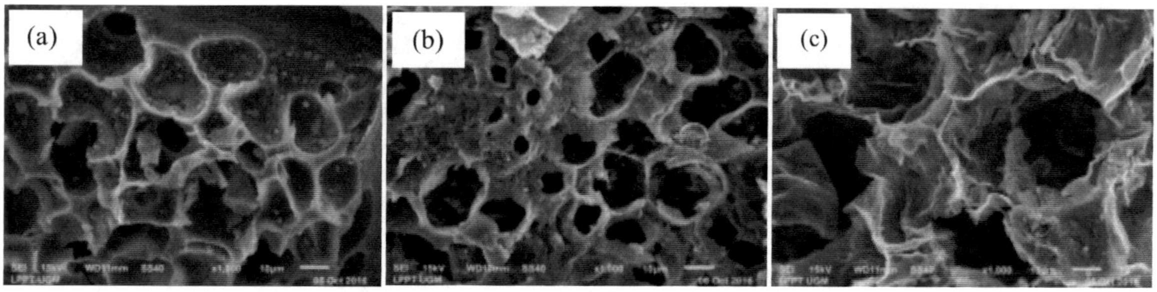

Figure 1. SEM micrographs of (a) char, (b) CAC and (c) MAC

To confirm the presence of functional groups which were capable to adsorb malachite green, FTIR analysis was conducted. Figure 2 describes the FTIR spectra of dried banana peel, char, and MAC. The wide absorption peak around 3400-3700 cm^{-1} with a maximum at 3430 cm^{-1} (Figure 2(a)) indicated the O-H stretching mode of hydroxyl groups derived from cellulose and lignin in the banana peel. While the absorption peak at 2924.56 cm^{-1} indicated the presence of the C-H group. The bands at 1642.28 cm^{-1} may be attributed to the C=C stretching vibration of olefinic groups, whereas the presence of C-OH groups derived from lignin was indicated by absorption peak at 1037.03 cm^{-1}. The existence of C-O-H twist bending vibration from cellulose was indicated by the presence of an absorption peak around 607 cm^{-1}. After the carbonization and activation process, the position and intensity of some peaks being changed (Figure 2(b) and 2(c)). Peak around 3400-3700 cm^{-1}, 1642 and 607 cm^{-1} became shallower which suggested the decrease of cellulose and lignin content.

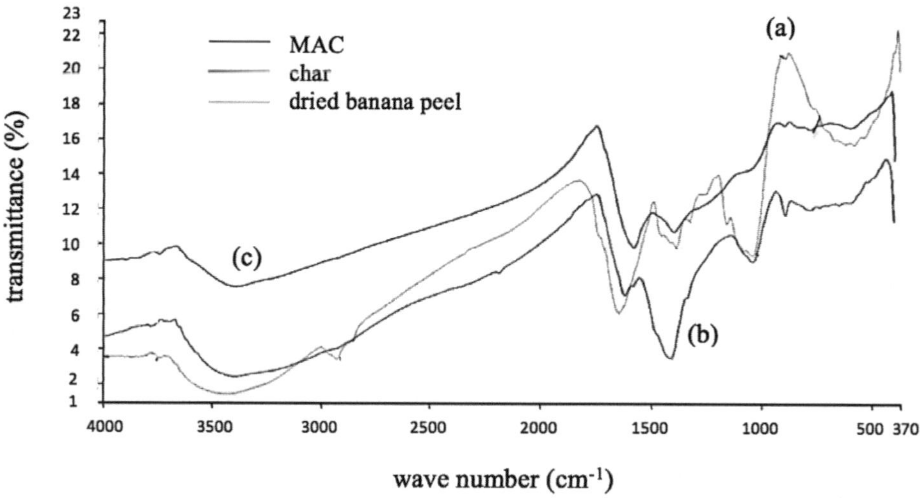

Figure 2. FTIR spectra of (a) dried banana peel, (b) char, and (c) MAC

BET surface area of CAC and MAC was 10.946 m^2/g and 187.664 m^2/g, respectively. It suggests that microwave heating increased surface area compared to conventional heating. Thus, there was a good agreement with the result of SEM micrographs which described that the pore size of CAC was smaller than that of MAC and several pores in CAC was blocked.

3.2. Adsorption of malachite green

The adsorption uptake of malachite green onto MAC linearly increased from pH 4 to 10, but it was nearly constant between pH 10 to 12 (Figure 3(a)). In the acid medium (i.e. presence of H^+), the phenolic groups contained in the MAC will be protonated to $R\text{-}O^+\text{-}H_2$. As a result, no ionic interaction between MAC and the malachite green cations. It led to a decrease in the amount of malachite green adsorbed. While, at higher solution pH, the negatively charged species start dominating in the MAC surface. As a result, the amount of malachite green adsorbed increased due to the electrostatic attraction between positively charged of malachite green and negatively charged of MAC. Between solution pH 10 to 12, adsorption was nearly constant due to the formation of carbinol complexes.

Figure 3(b) depicts that the quantity of malachite green adsorbed increased rapidly in the first 10 min and thereafter it gradually slows down till the equilibrium was reached. It may be due to a large number of vacant sites were available to the adsorption process during the initial stage.

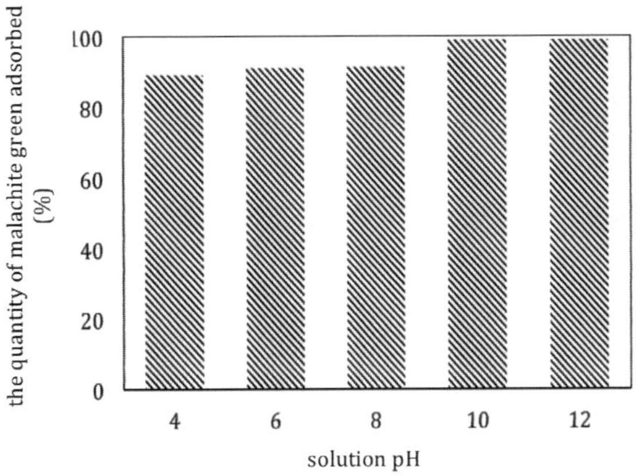

(a)

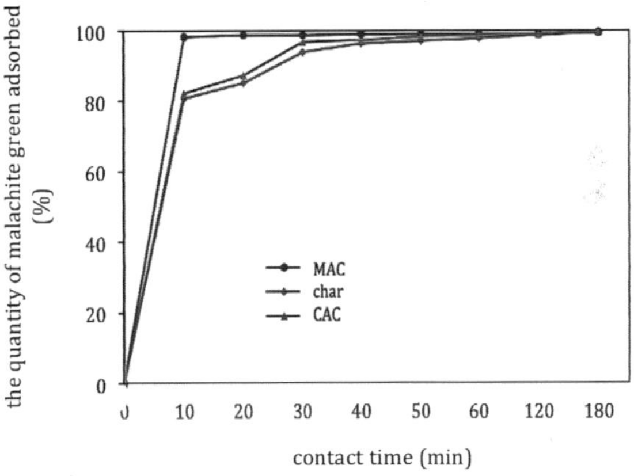

(b)

Figure 3. The effect of (a) pH and (b) contact time toward quantity of malachite green adsorbed

3.3. Adsorption isotherm study

Adsorption isotherm, that is used to optimize the design of adsorption system, describes the equilibrium of adsorbate molecules between liquid and solid (i.e. adsorbent) phase. It is characterized by specific constants that describe the affinity of adsorbent toward adsorbate molecules. In this study, two isotherm models including Langmuir and Freundlich were analyzed. Langmuir isotherm model is based on the assumption of monolayer adsorption with a homogeneous adsorbent surface. In this sense, adsorption sites on the adsorbent surface are assumed to be identical. The linear form of the Langmuir equation is described in Eq. (2) [13].

$$\frac{C_e}{q_e} = \frac{1}{q_m K_L} + \frac{1}{q_m} C_e \tag{2}$$

where q_m is the maximum adsorption capacity (mg/g) and K_L (L/mg) is Langmuir constant related to the affinity of the binding site. In contrast to the Langmuir model, the Freundlich model is an empirical equation based on the assumption of a heterogeneous adsorbent surface with a random distribution of adsorption heat over the adsorbent surface. The linear form of the Freundlich equation is described in Eq. (3) [13].

$$\log q_e = \log K_F + 1/n \log C_e \tag{3}$$

where K_F (L/g) is the Freundlich constant indicated the adsorption capacity while 1/n is a model parameter involving adsorption intensity. The value of isotherm parameters can be calculated from the slope and intercept of the graphed line in Figure 4. It shows that adsorption data be best fitted to the Freundlich isotherm model due to the value of R^2 for the Freundlich isotherm was closer to unity (R^2=0.965). It implied that the MAC surface was known to be highly heterogeneous with the value of K_F and n was 45.91 and 1.092, respectively.

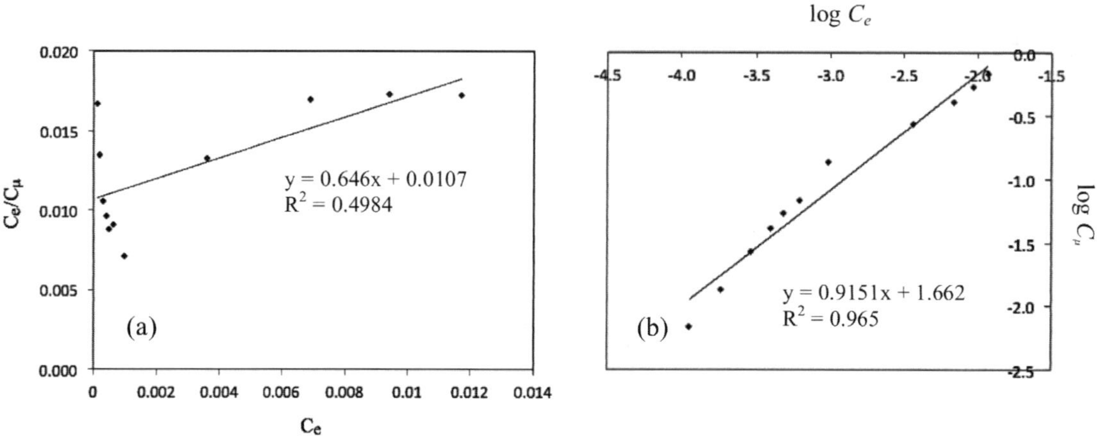

Figure 4. (a) Langmuir and (b) Freundlich isotherm plot for the malachite green adsorption by MAC

4. Conclusion

This study has demonstrated the suitability of kepok banana peel activated carbon prepared by microwave heating for malachite green adsorption. The optimum adsorption of the malachite green occurred at pH 10 for 180 min. Adsorption data be best fitted to the Freundlich isotherm model.

References

[1] Astuti W, Sulistyaningsih T, Maksiola M 2017 *Asian J. Chem.* **29** 133-138
[2] de Moura DC, Quiroz MA, da Silva DR, Saiazar R, Martinez-Huitle CA 2016 *Environmental Nanotechnology, Monitoring and Management* **5** 13-20
[3] Li H, Liu S, Zhao J, Feng N 2016 *Colloids and Surfaces A: Physicochemical and Engineering Aspects* **494** 222-227
[4] Zhang X, Lin Q, Luo S, Ruan K, Peng K 2018 *Appl. Surf. Sci.* **442** 322-331
[5] Chafidz A, Astuti W, Augustia V, Novira DT, Rofiah N 2018 *IOP Conf. Series: Earth and Environmental Science* **167** 012013
[6] Mahamad MN, Zaini MAA, Zakaria ZA 2015 *Int. Biodeterior. Biodegrad.* **102** 274-280
[7] Njoku VO, Foo KY, Asif M, Hameed BH 2014 *Chem. Eng. J.* **250** 198-204
[8] Spagnoli AA, Giannakoudakis DA, Bashkova S 2017 *J. Mol. Liq.* **229** 465-471
[9] Pallarés J, González-Cencerrado A, Arauzo I 2018 *Biomass Bioenergy* **115** 64-73
[10] Pezoti Jr. O, Cazetta AL, Souza IPAF, Bedin KC, Martins AC, Silva TL, Almeida VC 2014 *J. Ind. Eng. Chem.* **20(6)** 4401-4407
[11] Yorgun S, Yildiz D 2015 *J. Taiwan Ins. Chem. E.* **53** 122-131
[12] Ahmed MJ 2016 *J. Environ. Chem. Eng.* **4(1)** 89-99
[13] Altintig, E., Altundag, H., Tuzen, M, Sari, A 2017 *Chem. Eng. Res. Des.* **122** 151-63

IOP Conf. Series: Materials Science and Engineering **625** (2019) 012026 doi:10.1088/1757-899X/625/1/012026

The increase of biodiesel B20 oxidation stability using antioxidant obtained from soursop leaves extraction through microwave assisted extraction technique

Megawati[1], W Astuti[1], U M Arief[2], R R Ginting[3] and W Sammadikun[1]

[1] Chemical Engineering Department, Faculty of Engineering, Universitas Negeri Semarang, Indonesia
[2] Electrical Engineering Department, Faculty of Engineering, Universitas Negeri Semarang, Indonesia
[3] Department of Chemical Engineering, Graduate School of Engineering, Hiroshima University, Japan

Email: megawati@mail.unnes.ac.id

Abstract. The purpose of this research is to study the performance of soursop leaves extract as an antioxidant to inhibit biodiesel B20 oxidation. The antioxidant extraction was conducted with ethanol as the solvent (1:13 v/v) using microwave assisted extraction at powers of 300, 450, and 600 W for 2-10 min. In addition, the biodiesel oxidation was performed with air (2.3 L/min) at temperatures of 100, 110, and 120 °C for 0-70 min. The extraction results show that microwave power and extraction time strongly influence the extraction yield. The highest total phenolic concentrations of soursop leaves extract antioxidant obtained at 600 W and 8 min is about 0.608 mg/mL. From the oxidation results, it can also be seen that temperature and time influence the acid numbers of biodiesel B20. In 10 min, at 100, 110, and 120 °C, the acid numbers of biodiesel B20 are 0.09, 0.12, and 0.13 g NaOH/g. The mixture acid numbers of biodiesel B20 and soursop leaf extract are 0.126, 0.13, and 0.15 g NaOH/g. The optimization result shows that the activation energy of biodiesel B20 oxidation is 37.7 kJ/mol and the mixture of biodiesel B20 and soursop leaf extract is 41 kJ/mol. Therefore, it can be concluded that antioxidant additional is proven to inhibit the oxidation of biodiesel B20, as much as 47%.

1. Introduction

Soursop (*Annona muricata* L.) is the easiest type of plant to grow among other types of *Annona* in a warm and humid tropical climate [1]. The leaf themselves contain bioactive compounds, tannins, phenols, phytosterols, flavonoids, and saponins [2]. Flavonoids are phenolic compounds that can act as antioxidants, because they have a molecular structure that can transfer electrons to free radical molecules [3]. Conventional extraction methods like maceration and Soxhlation are the most common methods used to extract antioxidants, but they have several disadvantages; they need a large amount of solvents and long process, while the yield obtained is few [4]. A modern method which has been developed is microwave assisted extraction (MAE) [5]. MAE is an extraction method which utilizes microwave as a heating medium. This technology is suitable for extracting thermolabile antioxidants, because it has better temperature control than conventional heating methods. In addition, MAE also has several advantages [6]. This research used MAE to extract soursop leaf antioxidant efficiently and economically. The antioxidant obtained was used to increase the oxidation stability of biodiesel B20.

Indonesia has produced biodiesel B20 (20% v/v biodiesel + 80% v/v diesel fuel) as an alternative fuel for energy conservation. Biodiesel compounds produced from unsaturated fatty acids, especially methyl esters from vegetable and animal oils can cause oxidative degradation in biodiesel [7-9]. If biodiesel is exposed to high temperature, sunlight, metals and air, it will be oxidized to acid and alcohol [10]. The alcohol formed will increase the total acid and reduce the flash point, so it will be a problem on combustion engines, especially in the injection system [11]. During the oxidation process, acid numbers, kinematic viscosity, peroxide numbers, and ester levels in biodiesel deteriorate, so biodiesel oxidation needs to be inhibited [12]. In order to inhibit oxidation reactions, antioxidants need to be added [13].

A study on the performance of natural antioxidants from pepper extract, coffee leaf, bacupari leaf, and sage to inhibit biodiesel B100 oxidation has been conducted. Its performance was studied through an oxidation kinetics approach called pseudo-homogeneous first-order model. The effect of temperature on the oxidation kinetics was approached by the Arrhenius equation [14]. Meanwhile, in this research, the performance of antioxidant obtained from soursop leaf extraction on the oxidation kinetics will be evaluated based on the activation energy. This study aims to investigate the performance of soursop leaf extract as an antioxidant for biodiesel B20 through its oxidation kinetics using pseudo-homogeneous first-order model approach.

2. Materials and methods

2.1 Antioxidant extraction of soursop leaf using microwave assisted extraction

The materials used for this study were soursop leaf obtained from Perumahan Sapta Marga, Semarang. The chemicals used were 70% ethanol and gallic acid, folin ciocalteau reagent, Na_2CO_3, methanol, NaOH, and distilled water. Before being used, the soursop leaves were dried using an oven at 90 °C for 2 hours, blended using a blender until they became powder, and sifted using a 500 μm sieve. The soursop leaf extraction through MAE method was done by putting in 20 g of soursop leaf powder into a glass extractor and adding 70% ethanol as much as 260 mL. The extractions were conducted at 300, 450, and 600 W for 10 min. In every 2 min, 5 mL of the sample was taken in order to analyze its phenolic concentration. After the extraction was complete, the mixture was vacuum filtered to separate the filtrate and residue. The solvent in the filtrate was recovered by distillation at 78 °C until the extract volume was 20 mL. The extract obtained was purified again using an oven at a temperature of 100 °C to obtain pure extract.

The phenolic concentration analysis was conducted using UV-Vis spectrophotometer by adding 0.5 mL samples with 0.5 mL folin ciocalteau reagent (v/v) and 4 mL Na_2CO_3 7.5% (w/v). The solution was left for 60 min. Absorbance was determined at a wavelength of 765 nm. The phenolic concentration of antioxidant extract was calculated based on the absorbance produced using a calibration curve. In each power, total phenolic levels can be calculated from data on antioxidant phenolic concentrations obtained every 2 min and expressed as mg gallic acid equivalent (GAE)/g extract.

2.2 Oxidation of biodiesel B20

Biodiesel was obtained from PT. Pertamina. Before being used, physical and chemical properties of biodiesel B20 and the mixture of biodiesel B20 and soursop leaf extract were analyzed. The density analysis was conducted by Pyrex glass pycnometer (volume of 5 mL) while the kinematic viscosity analysis was done by viscometer (Stanhope-Seta - KV6 Viscometer Bath). In The first step of biodiesel oxidation, biodiesel B20 sample and antioxidant (5% v/v), were put into a flask and then heated and stirred at 600 rpm for 20 min using a speed regulator hotplate. After reaching 100 °C (being kept constant) air with a speed of 2.5 L/min was injected. Sampling was carried out in every 10 min for seven times. This treatment was repeated for different temperatures (110 and 120 °C). This oxidation was conducted for both biodiesel B20 and the mixture of biodiesel B20 and antioxidant.

2.3 Kinetic of biodiesel B20 oxidation

Oxidation conversion (x) can be calculated using equation (1), with AV_t = acid number when t (mg NaOH/g sample) and AV_0 = initial acid number.

$$x = \frac{AV_t - AV_0}{AV_0} \quad (1)$$

The performance of antioxidant was analysed through its oxidation reaction kinetics. Assumptions taken to solve the reaction rate were: 1) The reaction runs in batches with a fixed volume; 2) decrease in oxygen can be ignored because it is injected continuously [14,15]; 3) the kinetics were approached using pseudo-homogeneous first-order model.

3. Results and discussion

3.1 The influences of time and microwave power on antioxidant extraction of soursop leaves using microwave assisted extraction

The influence of extraction times (2-10 min) on phenolic concentration resulted from soursop leaf extraction using MAE can be seen in Table 1. The longer the extraction time, the higher the phenolic concentration will be. However, after reaching a certain time, the phenolic concentration decreases. The longer the extraction time, the more effective the breakdown or degradation of cell walls will be [16]. According to Ghafoor et al., during extraction, the solute will continue to dissolve in the solvent and stop when the solvent is saturated with the solute [17]. In addition, according to Chew et al., the longer the extraction time, the phenolic compounds from *Centella asiatica* extraction can be degraded [18]. It can be concluded that, at 600 W and 10 min, phenolic compounds extracted from soursop leaf have been damaged. In the processes of using microwave as the energy source, the power parameters represent temperatures. Mario et al. stated that some phenolic compounds are sensitive to high temperature changes, so they can be damaged [19]. Meanwhile, in this study, the effect of power can be studied. At 2 min, the higher the power (300, 450, and 600 W), the higher the phenolic concentration (10.01, 12.20, and 17.08 mg/mL, respectively) and at 8 min, the higher power, phenolic concentrations also increased (0.393, 0.443, and 0.603 mg/mL, respectively). It is obvious that antioxidant obtained from extraction is influenced by microwave power. Power shows the amount of heat as the driving force for destroying the pore cells of the leaf powder, so that antioxidants can diffuse out. Therefore, if the power is high, the antioxidants will quickly get out [20].

Table 1. Total phenolic concentration as well as phenolic content of soursop leaf extraction using microwave assisted extraction (operating condition: raw material weight of 20 g, raw material size of 500 μm, and 70% ethanol volume as solvent of 260 mL)

Time (min)	Concentration Phenolic (mg/mL)			Total Phenolic Content (mgGAE/g)		
	300 W	450 W	600 W	300 W	450 W	600 W
2	0.250	0.305	0.427	10.006	12.196	17.078
4	0.303	0.358	0.469	12.128	14.312	18.769
6	0.373	0.399	0.600	14.919	15.957	24.011
8	0.393	0.443	0.603	15.702	17.716	24.105
10	0.418	0.507	0.589	16.723	20.298	23.554

3.2 Physical and chemical properties of biodiesel B20 and the mixture of biodiesel B20 and soursop leaf extract antioxidant

The addition of antioxidants increases the density of biodiesel B20, because antioxidants have more density than biodiesel [21]. In this study, the density of biodiesel B20 obtained from PT Pertamina was 824 kg/m^{3}and the mixture of biodiesel B20 and soursop leaf extract antioxidant was 834 kg/m^3.

Meanwhile, the viscosity of biodiesel B20 was 2.92 mm^2/s and the mixture of biodiesel B20 and soursop leaf extract antioxidant was 2.89 mm^2/s. This is in line with a research conducted by Kivevele and Huan in which the addition of antioxidant additives can reduce the viscosity of biodiesel [22]. On top of that, the viscosity of biodiesel will increase due to oxidation reactions [8,23]. Thus, oxidation of biodiesel must indeed be prevented. In this study, biodiesel B20's acid numbers during the oxidation can be seen in Figure 1 and 2. It can be seen that the higher the temperatures, the higher the acid numbers will be. Soursop leaf extract antioxidant can prevent the oxidation of biodiesel B20 by 46.90%. It can also be said that antioxidant obtained from soursop leaf extraction is promising to be applied on a large scale in which biodiesel B20 oxidation kinetics data is needed.

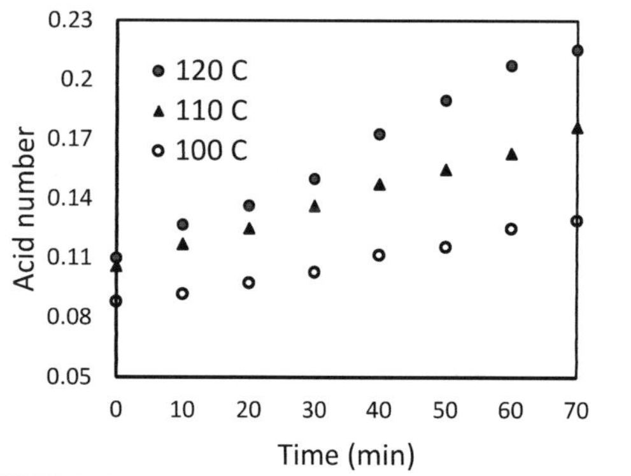

Figure 1. Acid numbers of biodiesel B20 during the oxidation process with air (operating conditions: rotation speed of 600 rpm and air velocity of 2.3 L/min).

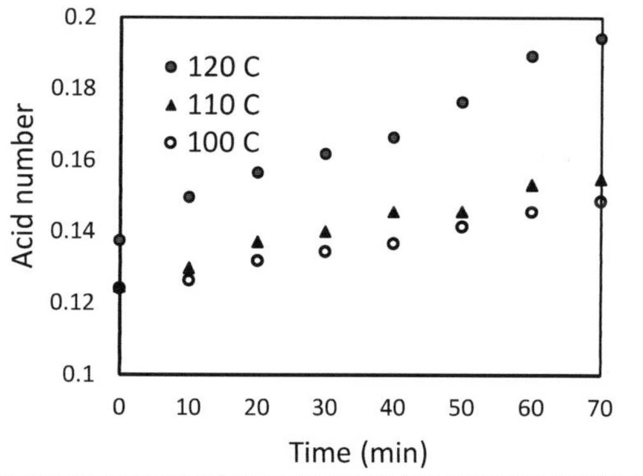

Figure 2. Acid numbers of the mixture of biodiesel B20 and soursop leaf extract antioxidant during the oxidation process with air (operating conditions: rotation speed of 600 rpm and air velocity of 2.3 L/min).

Oxidation kinetics are solved using the pseudo-homogeneous first-order reaction kinetics approach. The biodiesel oxidation rate constants (k_{bio}) are greater than those in the mixture of biodiesel and antioxidant (k_{mix}). That means that biodiesel oxidation is faster than the oxidation of the mixture of biodiesel and antioxidant. This shows that antioxidants can inhibit biodiesel oxidation. The comparison values of the oxidation rate constants (k_{mix}/k_{bio}) are as follows: 0.0026:0.0056; 0.0034:0.0076; 0.0051:0.0104, for 100, 110, and 120°C, respectively. The same result was also stated by Xin et al. and Gregorio et al. [14,15]. It is also seen that the reaction rate constant of each temperature follows the Arrhenius theory. From the results of linear regression, the activation energy values obtained were 37 and 41 kJ/mol for biodiesel B20 and the mixture of biodiesel B20 and soursop leaf extract antioxidant. From the increase in activation energy, it can be concluded that antioxidants play an important role in inhibiting oxidation reactions [14].

4. Conclusions

Soursop leaf extraction with 70% ethanol using microwave assisted extraction obtained a maximum yield of 33.98% (with total phenol of 24.10 mgGAE/g extract) at 600 W and 8 min. The extraction kinetics of soursop leaf were approached using the mass transfer from the particle's surface to the bulk of the solution, that can be expressed by second-order rate law. The soursop leaf extract antioxidant could inhibit biodiesel B20 oxidation by 47%. Biodiesel oxidation kinetics are strongly influenced by temperature with activation energy of 37 and 41 kJ/mol for biodiesel and the mixture of biodiesel and the soursop leaf extract antioxidant.

Acknowledgments

I acknowledge the support of the Ministry of Research, Technology, and Higher Education of the Republic Indonesia (Kemenristek-Dikti) 2019 in the form of an International Travel Grant, which

IOP Conf. Series: Materials Science and Engineering **625** (2019) 012026 doi:10.1088/1757-899X/625/1/012026

enabled me to attend this conference. This research was supported by Faculty of Engineering, Universitas Negeri Semarang (grant numbers 42.11.4/UN37/PPK.4.5/2018). We would also like to thank Handoyo, Gagah Muhammad Bungsu, and Muhammad Habib Fahd Rizky for their assistance in collecting the data and doing the oxidation kinetic modeling.

References

[1] Coria-Téllez A V, Montalvo-Gónzalez E, Yahia E M and Obledo-Vázquez E N 2018 *J. Chem.* **11** 662-691
[2] Baskar R, Rajeswari V and Kumar T S 2007 *J. Indian. Exp. Biol.* **45** 480–485
[3] Adeyemi D O and Komolafe O A 2009 *J. Afr. Trad.* **6** 62–69
[4] Megawati, Wulansarie R, Faiz M B, Adi S and Sammadikun W 2018 *AIP Conf. Proc.* **1941** 020016 1-7
[5] Farida R and Nisa F C 2015 *Jurnal Pangan dan Agroindustri* **2** 362-373
[6] Megawati and Murniyawati F 2015 *Jurnal Bahan Alam Terbarukan* **4** 14-20
[7] Domingos A K, Saad E B, Vechiatto W D, Wilhelm H M and Ramos L P 2007 *J. Braz. Chem. Soc.* **18** 416-423
[8] Knothe G 2007 *Fuel Proc. Technol.* **88** 669–677
[9] Anis S, Budiandono G N, Saputro D D, Zainul Z A 2018 *Jurnal Bahan Alam Terbarukan* **7** (in progress)
[10] Chan H W S 2005 *Autoxidation of Unsaturated Lipid*, ed H W S Chan (New York: Academic Press) p 1
[11] Sarin R, Sharma M, Sinharay S and Malhotra R K 2007 *Fuel* **86** 1365-1371
[12] Galvan D, Orives J R, Coppo R L, Silva E T, Angilelli K G and Borsato D 2013 *Ener. & Fuel* **27** 6866–6871
[13] Mittelbach M and Schober S 2003 *J. Am. Oil Chem. Soc.* **80** 817-823
[14] Gregorio A P H, Borsato D, Moreira I, Silva E T, Romagnoli E S and Spacino K R 2017 *Biofuels* **8** 1-8
[15] Xin J and Imahara H S 2009 *Fuel* **88** 282–286
[16] Handayani P A and Juniarti E R 2012 *Jurnal Bahan Alam Terbarukan* **1** 1–7
[17] Ghafoor K and Yuan H C 2009 *J. Korean Soc. Appl. Biol.* **52** 295-300
[18] Chew K K, Ng S Y, Thoo Y Y, Khoo M Z,Wan A W M, Ho C W 2011 *Internat. Food Res. J.* **18** 571-578
[19] Mário R M J, Alice V L, Nathalia R V D 2010 *The Open Chem. Eng. J. Brazil.* **4** 51-60
[20] Calinescu I, Ciuculescu C, Popescu M, Bajenaru S and Epure G 2001 *Romanian. Internat. Conf. Chem. Chem. Eng.* **12** 1-6
[21] Fattah I M R, Masjuki H H, Kalam M A, Mofijur M and Abedin M J 2014 *Ener. Convers. Manage.* **79** 265–72
[22] Kivevele T T and Huan Z 2013 *Ener. Tech.* **1** 537-543
[23] Fernandes D M, Serqueira D S, Portela F M, Assunção R M N, Munoz R A A and Terrones M G H 2012 *Fuel* **97** 658–661

IOP Conf. Series: Materials Science and Engineering **625** (2019) 012027 doi:10.1088/1757-899X/625/1/012027

Functional targeting nanomicelles with ph-triggered peptide activation for improved drug delivery in lung cancer

Yuehua Wang[1,2], Fei Jia[2,3], Chunyan Yue[2,3], Jian Sun1[*], Weizhi Wang[2*]

[1]School of Pharmaceutical Science and Technology, Health Science Platform, Tianjin University, Tianjin 300072, China;

[2]CAS Key Laboratory of Standardization and Measurement for Nanotechnology, CAS Key Laboratory for Biomedical Effects of Nanomaterials and Nanosafety, CAS Center for Excellence in Nanoscience, National Center for Nanoscience and Technology of China, Beijing 100190, P. R. China

[3]University of Chinese Academy of Sciences, Beijing 100049, P. R. China

Email: wangyh@nanoctr.cn

Abstract. Mild acidic environment and tumor angiogenesis are two typical features. The specific response toward lower pH may increase the targeting ability both for drug delivery. Herein, we present a responding self-assembled nanoparticle that is tumor angiogenesis targeting and pH triggered based on a novel recognition peptide DP. DP is pH-responsive as well as the affinity ligand of tumor marker EGFR. Efficient drug delivery and in vitro therapy in cell level could be accelerated in the conditions of EGFR overexpression and a mild acidic environment. We believed that such responsive materials will open a new way to turn on target imaging diagnostics and specific cancer therapies.

1. Introduction

Lung cancer, as one of the severest cancers all over the world, shows a rising trend in incidence rate for the past few years, bringing patients and their family great suffering and economic pressure.[1] However, traditional chemotherapeutic drugs are lack of specificity for swollen and painful cells, and it is difficult to achieve effective concentration inside the tumor.[2] While killing tumor cells, they will also cause damage to some normal tissues and cells in the body, and have disadvantages.[3] Hence, a series of tumor-targeting therapy have been studied currently. These studies show that there are a lot of differences in micro-environment between tumors and normal tissues, such as angiogenesis, permeability, and the extracellular pH differences, which can be used for drugs to recognize tumors specifically.[4] The pH in tumor microenvironment varies with site of tumors, as 6.0-7.0, while pH in blood and normal tissues is about 7.4.[5] All studies support that low pH promotes the development and tumor metastasis.[6] And high rate of glycolysis results in low pH at hypoxic tumor cells, which can be a target for tumor therapy.[7]

Herein, micelle is a new form of targeted drug delivery system and is a hot research object in the field of nanometer. It is a micelle formed with the hydrophilic end of amphiphilic polymer as the shell and the hydrophobic end as the core, which can selectively deliver drugs to the tumor site to play a therapeutic role without affecting the function of normal cells, tissues or organs, so as to improve the efficacy and reduce the toxic and side effects.[8] With a small particle size between 10-80 nm, the drug accumulates in tumor tissues through EPR effect to achieve passive targeting.[9] The critical micelle concentration was low and the dilution stability was good after administration in vivo. When the shell is connected with the specific ligand, the physiological effect of the drug can be improved by active targeting.[10] So, micelles target EGFR, which is highly expressing in non-small cell lung cancer, for targeting tumor tissues and lowering drug toxicity.

2. Results and discussion

2.1 Design of the self-assembled DENMEI

As shown in Scheme 1, the amphiphilic self-assembling nanostructures were first constructed. In this work, a switchable and functional nanoparticle system was established based on nanomicelles encapsulating Erlotinib (EI). The micelles were functionalized by one type of peptide DP (DP: CHLWEFDHA; Scheme 1a) with both pH-responsive and EGFR-targeting ability. This novel peptide shows high specificity toward EGFR in a mild acidic microenvironment. DENM with EI encapsulated in it ($DENM^{EI}$) exhibits high efficiency of targeting, release, and penetration in the tumor microenvironment toward EGFR in vitro (Scheme 1b). As an environment-triggered carrier, $DENM^{EI}$ could perform with the self-assembling ability that was of significance for cellular delivery.

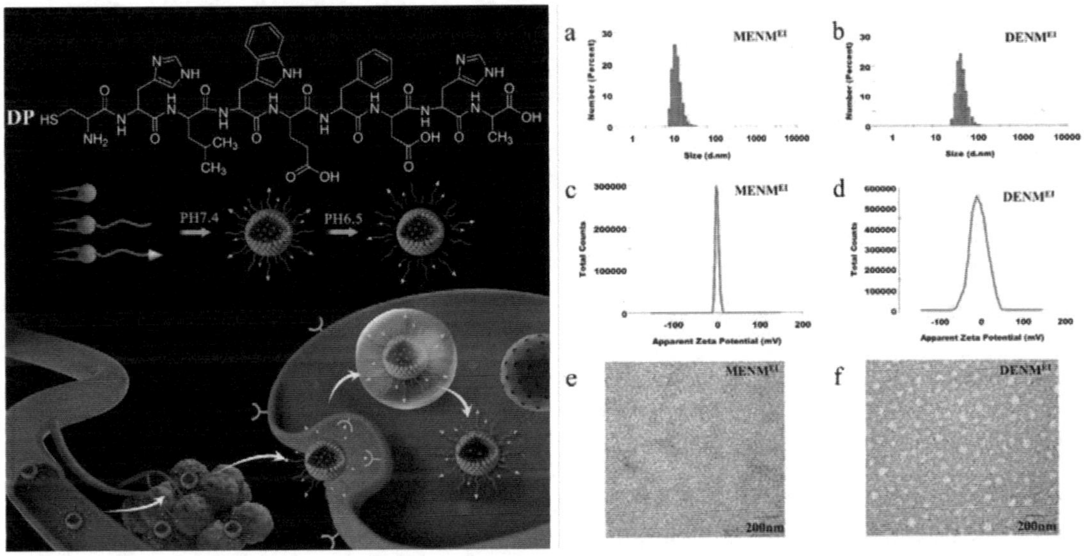

Figure 1. (Left) Schematic illustration of the preparation and application of $DENM^{EI}$. (Right) a. Size distribution of $MENM^{EI}$. b. Size distribution of $DENM^{EI}$. c. TEM graph of $MENM^{EI}$. d. TEM graph of $DENM^{EI}$.

2.2 Synthesis and characterization of $DENM^{EI}$ and $MENM^{EI}$.

The distribution of size, surface potential and the TEM images of formed micelles were presented in Figure 1. As shown in Figure 1a and b, the average diameter of micelles were determined by DLS to be 36.3 ± 3.85 nm (PDI = 0.283) for $MENM^{EI}$ and 67.4 ± 2.53 nm (PDI = 0.247) for $DENM^{EI}$ in acidic environment. When hydrophobic drug was further entrapped into the hydrophobic core, the micelle size was always larger than blank micelles. These might result from the stretch of PEG chains and hydrophilous peptide DP on the surface of micelle toward water phase. The change in the size explained that modification might be achieved. As shown in Figure 1c and d, zeta potential measurements revealed an overall neutral charge for both micelles, with a charge of -0.99 ± 0.17 mv for $MENM^{EI}$ and -5.76 ± 0.64 mv for $DENM^{EI}$, respectively. The absolute values of the ζ potential revealed that both $MENM^{EI}$ and $DENM^{EI}$ exhibited good dispersion stability in acidic environment, which was suitable for the subsequent experiment. Moreover, the morphologies of $MENM^{EI}$ and $DENM^{EI}$ were observed by transmission electron microscopy. As shown in Figure 1e and f, the morphology feature demonstrated that DENM^{EI} could self-assemble to form uniformly spherical nanoparticles.

2.3 The prepared nanomicelles were characterized by FTIR analysis.

The targeted peptide DP modified CMC micelle was prepared using MAl-PEG-NHS as a linker. PEG was assembled to CMC by the reaction between NHS and NH_2, following sulfhydryl group of peptides grafted to maleimide functional sites of PEG at pH 6.5. Peptide-free mPEG-CMC was chosen as a control, and both micelles were assessed by using FTIR (Fourier Transform infrared spectroscopy). As shown in figure 2, the FTIR spectra showed mPEG grafted carboxymethyl chitosan with characteristic peaks at 1102 cm-1(C-O-C stretching), and 2870 cm-1 (aliphatic CH2 stretching) belonging to mPEG. The characteristic peaks at 1622 cm-1 (N-H stretching vibration) and 1385 cm-1 (C-N stretching) also indicated the link of mPEG onto CMC by reacting to NH2 (1602 cm-1). In the IR spectrum of DP-PEG-CMC, a new peak at 1710 cm-1 might correspond to the C=O stretching of carboxyl groups belonging to peptides and the peak at 1344 cm-1 could attributed to the C-S stretching, indicating the grafting of peptides to PEG.

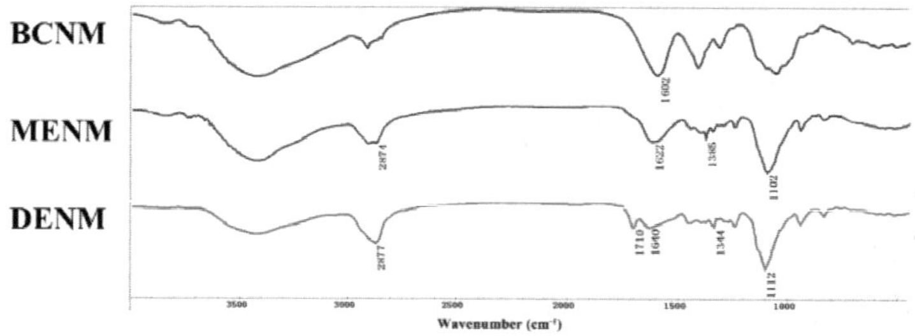

Figure 2. FTIR spectra of BCNM (control), MENM and DENM at pH 6.5.

2.4 Cell viability and drug release of the Erlotinib-loaded micelles on PC cells

MTT assays were performed to investigate the cytotoxicity of the MENM and DENM polymer and it loaded with Erlotinib. As shown in Figure 3a, the results proved that cell apoptosis of DENM^{EI} group at pH 6.5 was more significant than that of DENM^{EI} group at pH 7.4 and other groups with increasing concentrations, and especially at high concentrations. Furthermore, the DENM and MENM group showed no apparent difference at all concentrations in terms of cytotoxicity under the condition of pH 7.4 and pH 6.5. This might be caused by the better cellular uptake of the DP-linked DENM^{EI} complexes when applied to the EGFR-positive PC9 cells, and this was further supported by the results of the cellular uptake assays.

Controlled and sustained drug release is very important in drug delivery systems. During the study, the rate of drug release in both pH 6.5 and pH 7.4 was evaluated by dialysis method. Erlotinib (EI) was packaged on the above vector (DENM and MENM) for detection and identification. As shown in Figure 3b, the rate of Erlotinib release of DENM^{EI} at pH 7.4 was lower than at pH 6.5 (90% and 94%, respectively). At the same time, the release rate of MENM^{EI} at pH 7.4 was lower than that of the system at pH 6.5 (80% and 86%, respectively). When the pH is lowered from 7.4 to 6.5, the dissociation of Erlotinib increases due to the positive charge repulsion between the drug molecules. At the same time, DSPE-PEG_{2000} changes from the original negative charge into a positive charge and weakens the electrostatic attraction between Erlotinib and nanomicelle. Therefore, DENM^{EI} may be accelerated by endogenous lysosomes under acidic conditions after endocytosis by tumor cells.

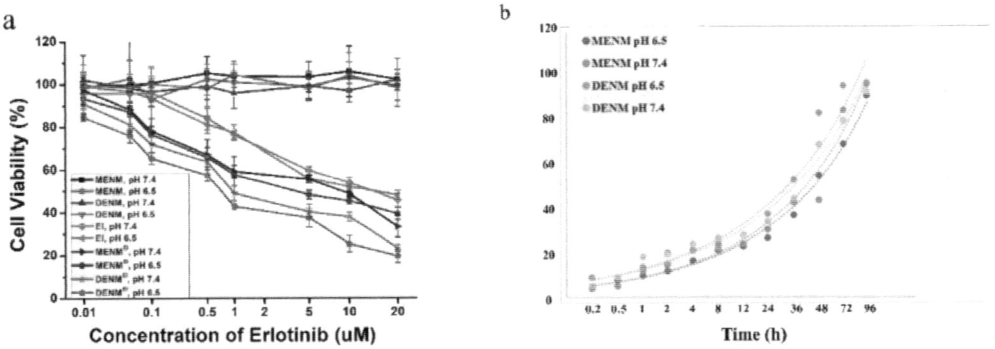

Figure 3. a. MTT assay for PC9 cells incubated with different Erlotinib formulations at different concentrations. b. *In vitro* release of Erlotinib from DENM^{EI} and MENM^{EI} at pH 7.4 and pH 6.5 (n=3).

2.5 Specific recognition of DENMEI at pH 6.5 in living cells

To evaluate the targeting delivery efficiency of DENM^{EI} toward EGFR positive cells at different pH environments, the human lung adenocarcinoma cell line PC9 was employed. As shown in Figure 4a, the nucleus of fluorescence signals were observed at the blue (Hoechst 33342, ex, 405 nm; em, 488 nm) and yellow (Dil; ex, 549 nm; em, 565 nm) channels in order to trace the interaction between nanomicelles and cells. The results indicated that PC9 cells treated with DENM^{EI} demonstrated much stronger fluorescent intensity than MENM at pH 6.5. Dil was successfully combined with the cell membrane. On the contrary, at pH 7.4, lower fluorescence signals were observed in PC9 cells incubated with DENM^{EI}. This revealed that DENM^{EI} is inactive at neutral condition. It manifested higher uptake efficiency of DENM^{EI} in the PC9 cell line at pH 6.5 since the overexpression of EGFR, which indicated that targeting binding achieved.

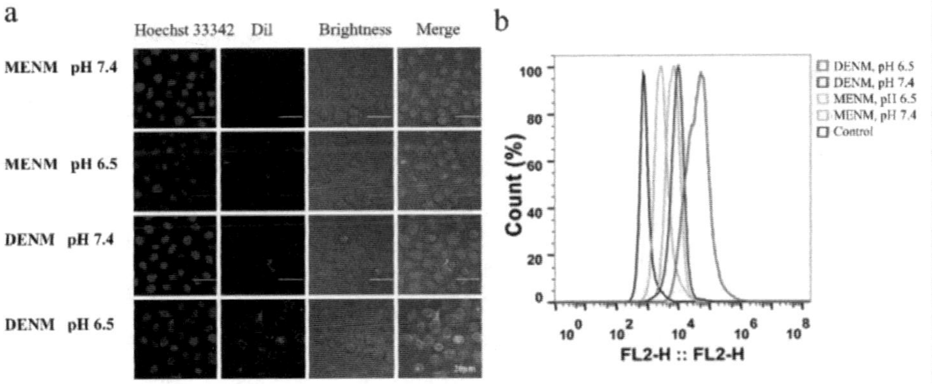

Figure 4. Characterization of the targeted recognition DENM^{EI} in cellular levels. a. CLSM images of PC9 cells incubated with MENM^{EI} or DENM^{EI} at pH 6.5 or pH 7.4 for 1 h (63×, oil-immersion objective). b. Flow cytometric analyses of the cellular uptake of MENM^{EI} or DENM^{EI} at pH 6.5 or pH 7.4.

3. Conclusion

This study systematically demonstrated a switchable micelle drug-delivery system, which functionalized by pH-triggered and EGFR-targeting peptide DP. The nanocarriers could enhance the antitumor effect at acidic tumor microenvironment and are expected to be potential candidates for lung cancer therapy. Our further work may pay more attention to the stability, specificity, and targeting affinity of the nanomicelles *in vivo*, in order to develop the enhanced synergy functional nanocarriers system, which could be prospectively applied in lung cancer therapy.

IOP Conf. Series: Materials Science and Engineering **625** (2019) 012027 doi:10.1088/1757-899X/625/1/012027

Acknowledgments
The authors are grateful for financial support from the National Natural Science Foundation of China (21775031, 81801766, 31870992), Beijing Natural Science Foundation (2172056 and L172035), Key Research Program of the Chinese Academy of Sciences (KFZD-SW-210).

References
[1] R. Mccall, M. Miles, P. Lascuna, B. Burney, Z. Patel, K. J. Sidoran, V. Sittaramane, J. Kocerha, D. A. Grossie and J. L. Sessler, *Chem Sci.* 2017 **8** 5918.
[2] L. Jianming, G. Caifang, Z. Ying, L. Chengli, W. Qi, H. Yongzhuo, Q. Jing, W. Jue, L. Weigen and W. Jianxin, *ACS Appl Mater Interfaces.* 2018 **10** 30201.
[3] P. Susanna, P. Jarmo, R. Derrick, K. Nina, R. Iiro, Y. H. Seppo, A. Kari, I. L. Weissman and S. Petri, *Proc Natl Acad Sci U S A* 2008 **105** 6620.
[4] J. Vandooren, G. Opdenakker, P. M. Loadman and D. R. Edwards, *Advanced Drug Delivery Reviews* 2016 **97** 144.
[5] R. Sluis, Van, Z. M. Bhujwalla, N. Raghunand, ., P. Ballesteros, ., J. Alvarez, ., S. Cerdán, ., J. P. Galons and R. J. Gillies, *Magn Reson Med.* 2015 **41** 743.
[6] J. S. Fang, R. D. Gillies and R. A. Gatenby, *Semin Cancer Biol.* 2008 **18** 330.
[7] F. Puglisi, D. Rea, M. A. Kroes and P. Pronzato, *Cancer Treat Rev.* 2016 **43** 36.
[8] F. Zhang, Q. Ni, O. Jacobson, S. Cheng, A. Liao, Z. Wang, Z. He, G. Yu, J. Song and Y. Ma, *Angew Chem Int Ed Engl.* 2018 **130** 7066.
[9] U. Evan, P. Thomas, L. Jonathan and G. Paul, *Adv Drug Deliv Rev* 2014 **72** 110.
[10] V. Uskoković, *J Biomed Nanotechnol.* 2013 **9** 1441.

IOP Conf. Series: Materials Science and Engineering **625** (2019) 012028 doi:10.1088/1757-899X/625/1/012028

Microstructure and properties of superconducting, ferromagnetic and hybrid nanowire networks of $La_{1.85}Sr_{0.15}CuO_4$ and $La_{0.5}Sr_{0.5}MnO_3$

M. R. Koblischka[1,2], A. Koblischka-Veneva[1,2], X. L. Zeng[1], T. Karwoth[1]

[1] Experimental Physics, Saarland University, P.O.Box 151150, 66041 Saarbrücken, Germany
[2] Superconducting Materials Laboratory, Department of Materials Science and Engineering, Shibaura Institute of Technology, Tokyo 135-8548, Japan

E-mail: miko@shibaura-it.ac.jp

Abstract. We have successfully fabricated $La_{1-x}Sr_xMnO_3$ (LSMO) nanowires with various x level, and $La_{1.85}Sr_{0.15}CuO_4$ (LSCO) nanowires/nanoribbons via electrospinning. The colossal magnetoresistance (CMR) of the LSMO nanowire networks have been investigated. and the T_c of the LSCO nanowires and nanoribbons are around 19.2 K and 29.3 K respectively. Furthermore, we have established a LSCO/LSMO nanowire hybrid system. From obervation by scanning electron microscopy, the average diameter of the nanowires is around 220 nm and the average length can reach over 50 μmm. The randomly aligned LSCO and LSMO nanowires show numerous connections and form a complicated hybrid network system. The nanowires are polycrystalline with the grain size of ~30 nm as confirmed by transmission electron microscopy and EBSD. According to four-probe electrical transportation measurements, the superconductivity of the hybrid sample is suppressed and an anti-magnetoresistance effect is observed. SQUID measurements of $M(T)$ and $M(H)$ were carried out as well, revealing the soft magnetic character of the nanowires.

1. Introduction

Using the electrospinning technique [1, 2] it is possible to prepare ceramic superconducting and ferromagnetic nanowire networks from sol-gel type precursors [3–6]. In previous works, we could fabricate ferromagnetic $La_{1-x}Sr_xMnO_3$ (LSMO) nanowires with various doping levels x [7] and superconducting $La_{1.85}Sr_{0.15}CuO_4$ (LSCO) nanowires [8]. LSCO was found to contain two different shapes of the nanoobjects: nanowires and nanoribbons. The latter ones are wide, but very thin nanowires which consist of practically one single layer of LSCO grains. Based on this experience it is natural to attempt the fabrication of a hybrid LSMO/LSCO nanowire network, and to to study the resulting microstructures and physical properties like it was done in the literature for multilayered thin films or mixed powder samples of the two components. As the unit cells of the two materials are very similar to each other, the formation of multilayers does not introduce too much strain [9–14]. In the present contribution, we compare the microstructures of the two components and of the hybrid nanowire networks and discuss the resulting physical properties.

2. Experimental procedures

2.1. Electrospinning

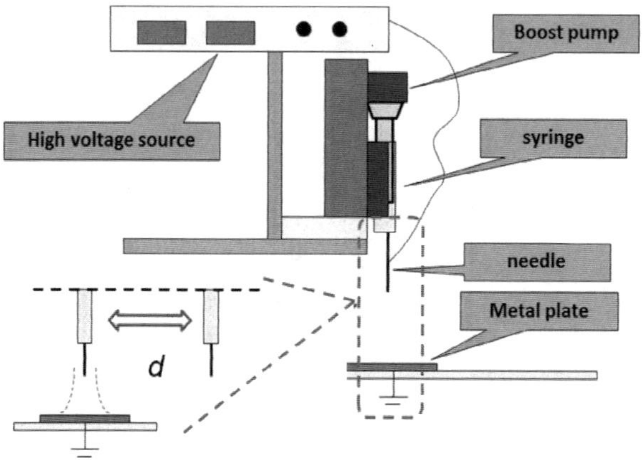

Figure 1. Schematic drawing of the electrospinning apparatus.The blow-up shows the situation when using two orifices to produce hybrid materials from two different sources. The distance *d* between the two orifices must be set properly to avoid interference effects.

Figure 1 presents a schematic drawing of the electrospinning instrument. The precursor is pushed by the microspeed boost pump from the syringe to the bottom of the needle. The precursor is dragged by the electric field between the needle and the grounding area, transforming the original droplets to nanofibers. This method is commonly used in the literature to prepare polymer nanofibers [1,2], and recently, also ceramic materials were demonstrated to be prepared using this technique [3–6]. The blow-up in Fig. 1 gives details about the use of two orifices for the spinning process of two component. Controlling the distance between two orifices as shown in the blue frame, it is possible to fabricate different materials simultaneously without interference with each other.

The electrospinning precursor is prepared by dissolving La, Sr, and Mn acetates in PVA (high molecular weight polyvinyl alcohol). The same process is carried out for LSCO using La, Sr, and Cu acetates. The PVA is slowly added to the acetate solution with a mass ratio of 2.5:1.5. This solution is stirred at 80 °C for 2 h, and then spun into cohering nanofibers by electrospinning. To remove the organic compounds and to form the desired ceramic phases, the sample is subsequently heat treated in a lab furnace. The reaction temperature is chosen to be 700 °C (1 h), which is close to the optimal temperature for LSCO. An additional oxygenation process is required to obtain the correct phase composition. Further details about the electrospinning process of these nanowires are given elsewhere [8, 15]. The constituent phases of the hybrid nanowire network samples were determined by means of a high-resolution automated RINT2200 X-ray powder diffractometer, using Cu-K^{α} radiation generated at 40 kV and a current of 40 mA.

2.2. Sample characterization

SEM/EDX imaging was carried out using a JEOL 7000 F SEM microscope operating at 15 kV. This microscope is further equipped with a TSL OIM analysis unt [16] for orientation imaging using the EBSD technique. More details about the EBSD measurements on such nanowire network samples can be found in [17]. TEM investigations were performed using a JEOL JSM-2011 transmission electron microscope operating at 200 KV with a LaB_6 cathode. For TEM and

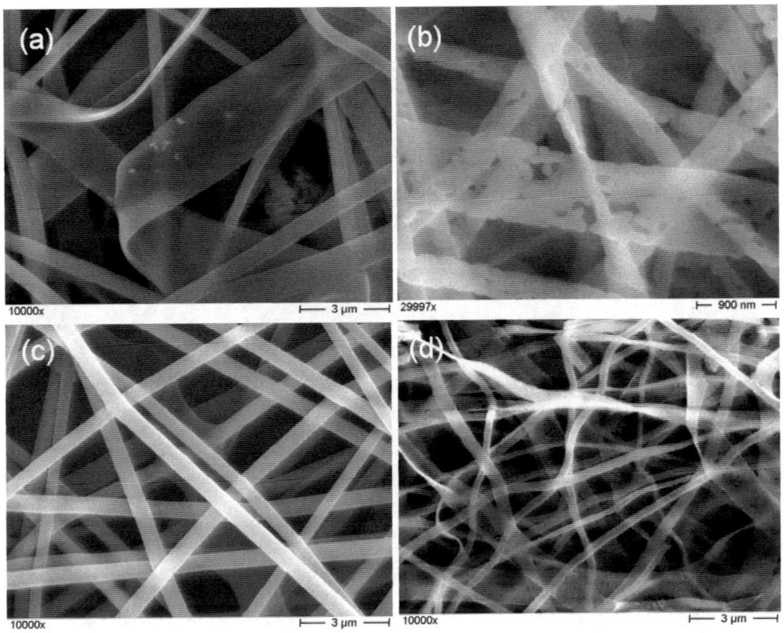

Figure 2. (a,c) SEM images of the LSCO nanoribbons. (a) as-spun, and (c) after the heat treatment. (b,d) SEM images of the LSCO nanowires, (c) as-spun, and (d) after the heat treatment. The as-spun polymer nanofibers are homogeneously smooth, whereas the heat-treated nanowires reveal the polycrystalline structure including voids.

t-EBSD investigations, several nanowire pieces separated from the as-grown nanowire fabric by ultrasound treatment and were placed on a carbon-coated TEM aluminium grid. This enabled to search nanowire regions which were thin enough to enable the transmission mode. No additional treatment of the sample surface was carried out.

The magnetization of the nanowire networks were measured using a superconducting quantum interference device (SQUID) magnetometer (Quantum Design MPMS3) with ± 7 T magnetic field applied perpendicular to the sample surface, using a piece of the nanowire network with a size of ~ 15 mm^2. The magnetoresistance was measured in a 10/12 T bath cryostat (Oxford Instruments Teslatron) with a Keithley source meter (model 2400) as a current source, and the voltage was recorded using a Keithley 2001 voltmeter.

3. Results and discussion

3.1. Microstructures

In our first experiments on LSCO nanowires, we found that two different types of nanoobjects result: nanowires and nanoribbons [8]. The nanoribbons are very thin, but wide, whereas the nanowires have a round cross section. Figure 2 presents SEM images of the LSCO nanowires and nanoribbons in the as-spun state directly after the electrospinning process (a,c) and after the heat treatment (b,d).

The as-spun polymer nanofibers are completely smooth and homogeneous. In contrast, the heat-treated nanowires and -ribbons show the resulting granular structure after the burn-off of the polymer and the ceramic grain growth. The images of the nanoribbons further show that the ribbons are so thin, being transparent to the electron beam. These structures were maintained during the whole thermal treatment. The average diameter of the nanowires was determined to about 234 nm. The flat nanoribbons showed widths which are ~ 1 μm, but the corresponding

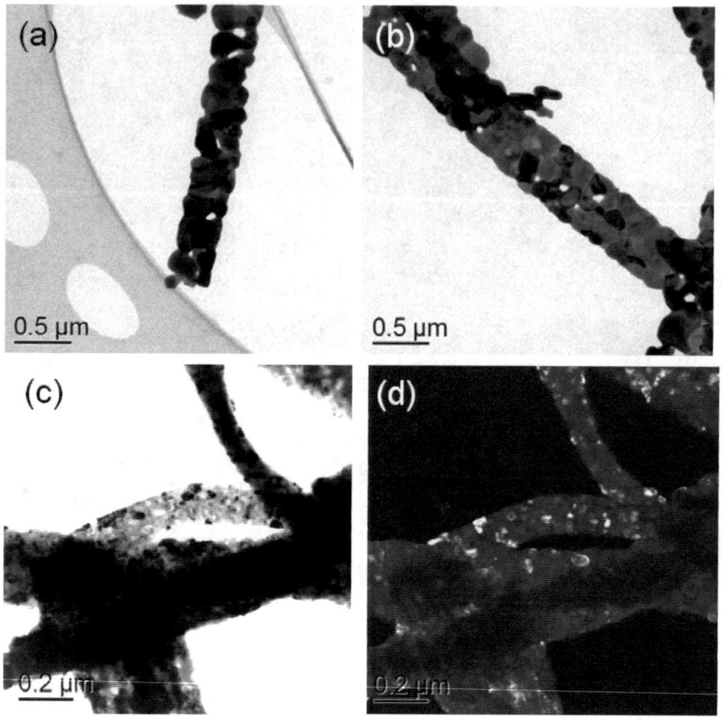

Figure 3. LSCO TEM images. (a) and (b) show nanoribbons, which are transparent to the electron beam. Images (c) and (d) are bright field and dark field images to reveal the details of the LSCO grain arrangement.

thicknesses are only about 60 – 80 nm. The total lengths of these fibers were found to be above 50 μmm; most of them were in the 100 μm range.

In Fig. 3, we present TEM images of the LSCO nanowires/nanoribbons. The resulting images reveal clearly that the samples are polycrystalline. The nanoribbons (a,b) consist of only one layer of grains, and show the presence of holes between the grains. The nanowires, shown here in bright and dark field images (c,d) to reveal the grain arrangement, exhibit the same shape of the grains as found at the nanoribbons, with well defined grain boundaries between the LSCO grains. The grain size statistics obtained from several TEM images shows a broad variation and the average grain size is determined to be about 110 nm. One should notice here that the grain size of the nanoribbons is equal to the thickness. This indicates that the nanoribbons are formed by individual LSCO grains, linked to each other in a single layer. For the LSMO samples, we only observed the presence of nanowires in all experiments carried out [7]. The resulting arrangement and shape of the LSMO grains is, however, very similar to the one of LSCO as expected from the similarity of the crystal structures.

Figure 4 presents a TEM image of the LSMO/LSCO hybrid sample (a). From such images, it is not possible to distinguish between the two components, therefore, we have applied EDX mapping to image the spatial distribution of the Mn and Cu ions [18]. As result, we find that the individual nanowires are either LSCO or LSMO, and no intermix has occured along the nanowires. This can be expected from the fact that first the precursor-containing polymer nanowires were formed by electrospinning, and the process to build-up the grains takes place only within a given wire. The situation may, however, be different in the interconnets between the wires, which are numerous in the resulting nanowire network. Such a situation is depicted in Fig. 3 (c,d). Figure 4 (b) presents the result of the X-ray analysis of the hybrid sample, and

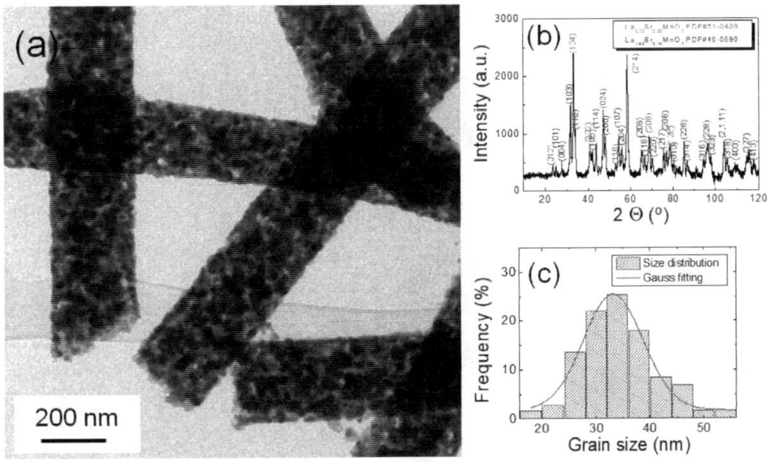

Figure 4. TEM image (a) of several individual nanowires from a LSMO/LSCO hybrid sample, revealing the polycrystalline, randomly-oriented grai arrangement. EDX mapping on these nanowires demonstrated that LSCO and LSMO nanowires are well separated. (b) shows the result of the X-ray analysis, and (c) gives the grain size distribution of the hybrid sample.

(c) gives the grain size statistics. The Gauss fitting of the data yields an average grain size of 33.2 nm.

Figure 5 presents the result of an EBSD measurement of the grain orientation performed on a LSMO nanowire. For this purpose, we followed the approach described in Ref. [17], using the transmission-EBSD technique on a single nanowire piece, which is thin enough. In (a), the measured section of the LSMO nanowire is shown (green box); the image (b) gives the EBSD-orientation mapping in [001]-direction (perpendicular to the surface). The color code for the orientation mapping is given in the stereographic triangle. The mapping reveals the polycrystalline character, together with the numerous high-angle grain boundaries. The plots (c) and (d) give the distribution of the grain shape aspect ratio, and the EBSD-determined grain size distribution, which is even smaller (~20 nm) as compared to the mean TEM value (33.2 nm). Thus, the EBSD analysis provides important additional information important for the evaluation of the MR data on such a sample.

3.2. Physical properties

In Fig. 6, we present the results of magnetization measurements on the nanowire networks. To obtain large enough magnetic signal, a certain amount of material is required for these measurements. In Fig. 6 (a), a $M(T)$ measurement on the LSCO nanowires/nanoribbons is presented, measured in field-cooling (FC) and zero field-cooling (ZFC) modes with an applied field of 1 mT. Here, we find two distinct superconducting transitions, T_c, at 29.3 K and at 13.9 K, respectively. We ascribe the two T_c to the contributions of the nanowires and nanoribbons; the superconducting phase being better developed in the nanoribbons [8].

In Figs. 7 (a,b) the resistance versus temperature curves measured in various applied magnetic fields (0 T, 2.5 T, 5.0 T and 7.5 T, perpendicular to the sample surface) is presented in the temperature range 2 K $\leq T \leq$ 300 K for the LSMO/LSCO hybrid sample (a) and the pure LSMO sample (b). The resistance of the LSMO/LSCO hybrid sample is obviously higher than that of the pure sample, showing the semiconducting behavior of LSCO in the normal regime. A comparison of the $R(T)$-data of the hybrid sample with the pure LSMO directly reveals an inversed field dependence.

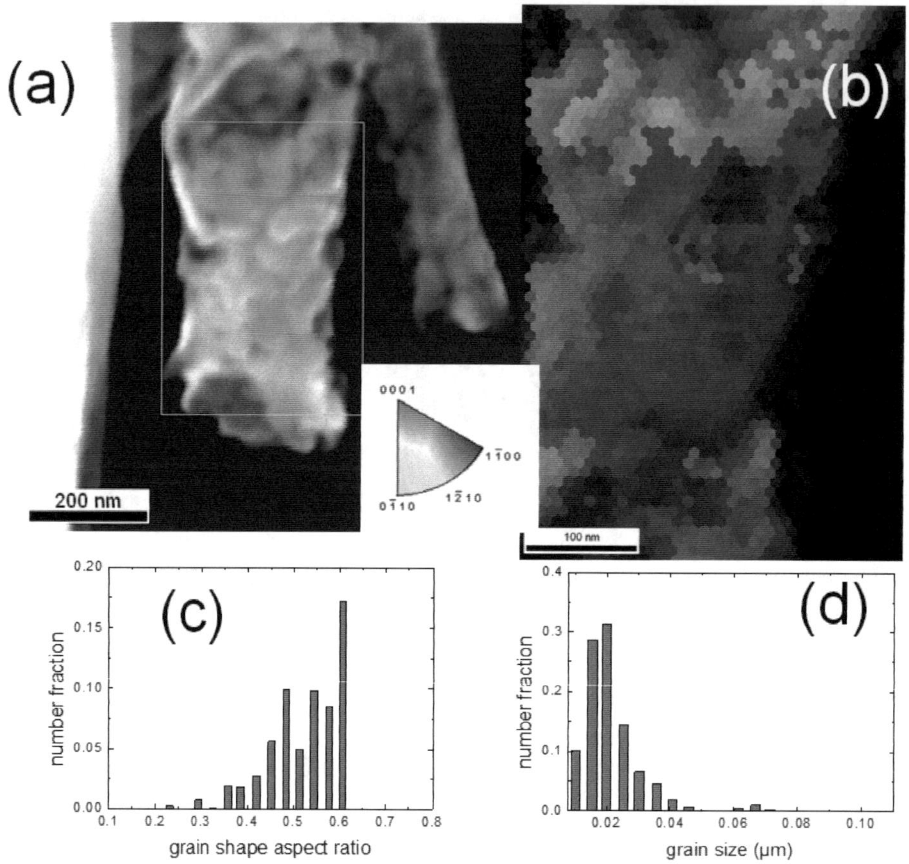

Figure 5. EBSD analysis of a LSMO nanowire. (a) SEM image of the selected area and (b) orientation mapping in [001]-direction. The color code for the mapping is given in the stereographic triangle, The plot (c) gives distribution of the grain shape aspect ratio and (d) grain size distribution.

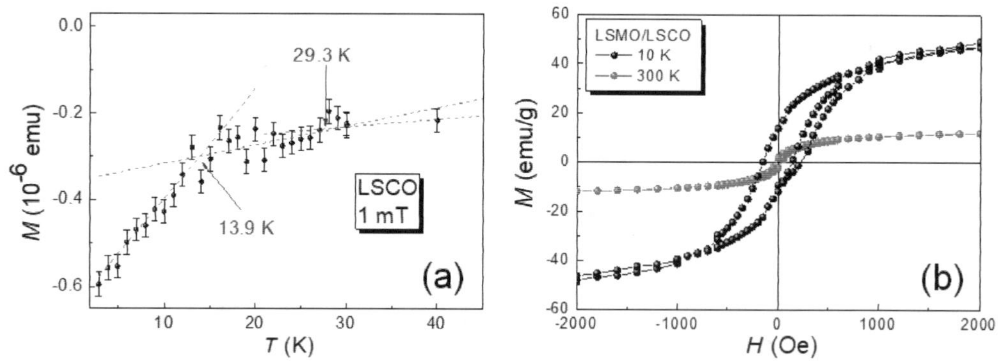

Figure 6. Magnetization measurements on the nanowire network samples. (a) gives an $M(T)$-measurement to demonstrate the onset of superconductivity in the LSCO nanowires. Two transititions can be seen – one at 29.3 K, the other at 13.9 K. (b) presents $M(H)$-loops at 300 K and at 10 K, showing weak ferromagnetic behavior.

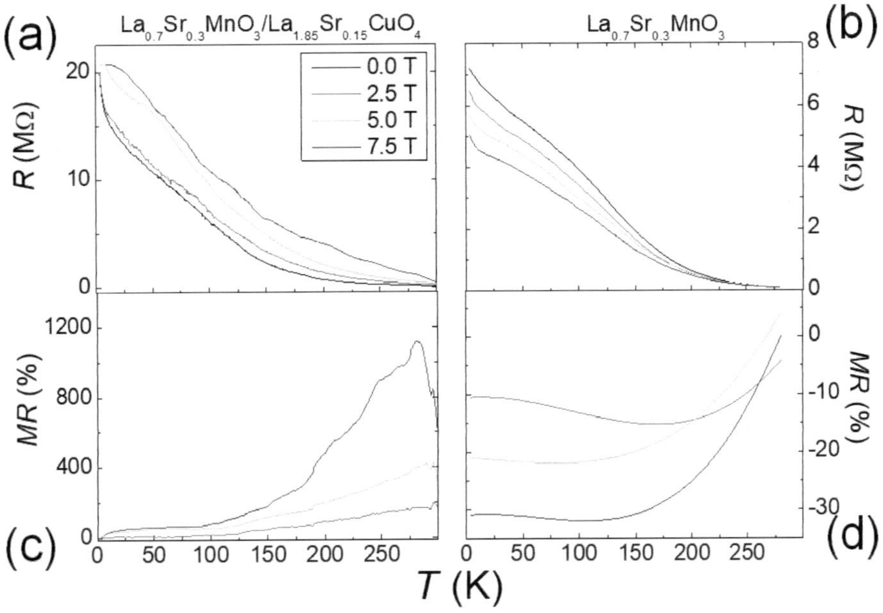

Figure 7. Resistance-temperature $R(T)$ diagram and the MR ratio for the LSMO/LSCO hybrid sample (left) and the pure LSMO sample (right) as function of temperature.

Figures 7 (c,d) give the MR ratio for both types of samples. Magnetoresistance (MR) plots are calculated from the resistance data obtained by sweep measurements at different field strengths using the relation $\mathrm{MR}[\%] = (R_H(T) - R_0(T))/R_0(T)$. From the MR plots of the LSMO/LSCO hybrid sample and the pure LSMO one, we can draw several interesting conclusions: (i) At low temperatures, the MR effect of the hybrid sample is clearly suppressed. (ii) The field dependence of the MR ratio is inversed, indicating a *positive* MR for the hybrid sample and a *negative* one for the pure LSMO sample as described previously in [7]. (iii) A large MR of ∼1120% is obtained in the hybrid sample at a temperature of 280 K and 7.5 T applied field. This maximum MR is strongly decreasing with lowering the applied field. These remarkable results may have different origins, partly induced by the nanowire structure and the small grain sizes of the constituents. The grain size of both LSMO and LSCO, together with the small nanowire diameter of ∼220 nm and the numerous interconnects between the individual nanowires in the nanowire network fabrics, provide a large number of interfaces, which the currents have to pass through. Both the grain boundaries and the interconnects between the wires lead to a scatter of the electrons, thus leading to a strong MR effect. The grain size effect on the MR properties was also previously discussed in [19, 20]. The grain size of the present nanowires is even slightly smaller as presented there. Thus, especially the high-field MR effect of the pure LSMO nanowires is clearly increased in our case. The role of superconductivity in the hybrid sample is also an important one. Even though our magnetic measurements could not reveal a superconducting diamagnetic contribution directly, the character of the interfaces between the nanowires is clearly changed. The superconducting signal may only be well developed within the LSCO grains, and there are not many interconnects between the wires as in these places an exchange between the two components becomes possible, and Mn ions may replace the Cu ions, leading to non-superconducting interconnects. The nature of these interconnects, therefore, determines the MR properties.

IOP Conf. Series: Materials Science and Engineering **625** (2019) 012028 doi:10.1088/1757-899X/625/1/012028

4. Conclusions

We have presented the fabrication and characterization of LSCO, LSMO and LSMO/LSCO nanowire hybrid network samples by electrospinning. The pure components are superconducting or ferromagnetic, but the hybrid sample exbibits interesting properties due the combination of superconductivity and ferromagnetism and the resulting character of the interconnects between the individual nanowires.

Acknowledgments

We thank J. Schmauch (Saarland University, group Prof. Birringer) for technical assistance and Prof. V. Presser for the use of the electrospinning apparatus. This work is supported by DFG grant Ko2323/8, which is gratefully acknowledged.

References

[1] Z. M. Huang, Y. Z. Zhang, M. Kotaki, S. Ramakrishna, "A review on polymer nanofibers by electrospinning and their applications", *Composites Science Technol.* **63**, 2223 (2003).

[2] D. Li, Y. N. Xia, "Electrospinning of nanofibers: Reinventing the wheel?", *Adv. Mater.* **16**, 1151(2004).

[3] Li D, McCann J T and Xia Y N 2006 Electrospinning: A simple and versatile technique for producing ceramic nanofibers and nanotubes. *J Am Ceram Soc* **89** 1861

[4] Wu H, Pan W, Lin D and Li H 2012 Electrospinning of ceramic nanofibers: Fabrication, assembly and applications. *J Adv Ceramics* **1** 2

[5] Li D, Herricks T and Xia Y N 2003 Magnetic nanofibers of nickel ferrite prepared by electrospinning. *Appl Phys Lett* **83** 4586

[6] Yensano R, Pinitsoontorn S, Amornkitbamrung V and Maensiri S 2014 Fabrication and magnetic properties of electrospun $La_{0.7}Sr_{0.3}MnO_3$ nanostructures. *J Supercond Novel Mag* **27** 1553

[7] Karwoth T, Zeng X L, Koblischka M R, Hartmann U, Chang C, Hauet T and Li J M 2019 Magnetoresistance and structural characterization of electrospun $La_{1-x}Sr_xMnO_3$ nanowire networks. *Solid State Commun* **290** 37

[8] Zeng X L, Koblischka M R and Hartmann U 2015 Synthesis and characterization of electrospun superconducting (La,Sr)CuO_4 nanowires and nanoribbons. *Mat. Res. Express* **2** 095022

[9] Liu J M, Huang Q, Li J, Ong C K and Lin Z G 1999 Pulseld laser deposition of perovskite $La_{0.7}Sr_{0.3}MnO_3/La_{0.7}Sr_{0.3}CoO_3$ multilayers and their magnetotransport properties. *Appl Phys A* **69**(suppl.) S663

[10] Li B, Chopdekar R V, N'Diaye A T, Mehta A, Paige Byers J, Browning N D, Arenholz E and Takamura Y 2016 Tuning interfacial exchange interactions via electronic reconstruction in transition-metal oxide heterostructures. *Appl Phys Lett* **109** 152401

[11] Yan C H, Xu Z G, Zhu T, Wang Z M, Cheng F X, Huang Y H and Liao C S 2000 A large low field colossal magnetoresistance in the $La_{0.7}Sr_{0.3}MnO_3$ and $CoFe_2O_4$ combined system. *J Appl Phys* **87** 5588

[12] Li X H, Huang Y H , Wang Z M and Yan C H 2002Tuning between negative and positive magnetoresistance in $(La_{0.7}Sr_{0.3}MnO_3)_{1-x}(La_{1.85}Sr_{0.15}CuO_4)_x$ composites. *Appl Phys Lett* **81** 307

[13] Habermeier H U, Cristiani G, Kremer R K, Lebedev O and van Tendeloo G 2001 Cuprate/manganite superlattices: A model system for a bulk ferromagnetic superconductor. *Physica C* **364-365** 298

[14] Yao X, Jin Y, Li M, Li Zh, Cao G, Cao Sh and Zhang J 2011 Coexistence of superconductivity and ferromagnetism in $La_{1.85}Sr_{0.15}CuO_4$-$La_{2/3}Sr_{1/3}MnO_3$ matrix composites. *J Alloy Compounds* **509** 5472

[15] Zeng X L, Koblischka M R, Karwoth T, Hauet T and Hartmann U 2017 Preparation of granular Bi-2212 nanowires by electrospinning. *Supercond Sci Technol* **30** 035014

[16] TexSEM Laboratories (TSL). Orientation Imaging Microscopy Software V4.1, User Manual, TexSEM laboratories, (TSL), Draper, UT, 2004.

[17] Koblischka-Veneva A, Koblischka M R, Zeng X L, Schmauch J and Hartmann U 2018 TEM and electron backscatter diffraction analysis (EBSD) on superconducting nanowires. *J Phys Conf Ser* **1054** 012005

[18] Zeng X L, Koblischka M R, Karwoth T and Hartmann U 2019 Properties of $La_{1.85}Sr_{0.15}CuO_4/La_{0.7}Sr_{0.3}MnO_3$ hybride nanowire networks prepared by electrospinning. *J Magn Magn Mater* **475** 741

[19] Kar S, Sarkar J, Ghosh B and Raychaudhuri A K 2007 Effect of grain boundaries on the local electronic transport in nanostructured films of colossal magnetoresistive manganites. *J Nanosci Nanotech* **7** 2051

[20] Balcells Ll, Fontcuberta J, Martinez B and Obradors X 1998 High field magnetoresistance at interfaces in manganese perovskites. *Phys Rev B* **58** R14697

IOP Conf. Series: Materials Science and Engineering **625** (2019) 012029 doi:10.1088/1757-899X/625/1/012029

Study of delignification process and crystallinity index on lignocellulose components of corn cob in different pretreatments: a combination of pretreatment (ionic choline acetate and NaOH) and NaOH pretreatment

U. Rofiqah[1,*], A. Safitri[1] and Fadhilah[1]

[1]Chemical Engineering Department, Universitas Islam Indonesia
Jalan Kaliurang km 14,5 Yogyakarta 55584, Indonesia

*E-mail: umirofiqah@uii.ac.id

Abstract. This study studied the effect of combination pretreatment (ionic liquid choline acetate and NaOH) and pretreatment of NaOH in the process of delignification and lignocellulose crystallinity index. This process was carried out to obtain a good lignocellulose biomass, which has a low lignin content and a high crystallinity index to be prepared as a raw material for making bioethanol. Low lignin content and high index of crystallinity can influence the success of the enzymatic hydrolysis process and of course bioethanol production. In the combination of pretreatment, the highest lignin removal was found from process with 5% of NaOH as much as 41.63% but the cristalinity index was found from proces with 3% of NaOH as much as 46.33. While, In the NaOH pretreatment, the highest lignin removal and cristalinity index was found from process with 5% of NaOH as much as 7.63% and 3.87, respectively.

1. Introduction

Corn is a dicotyledonous plant which is a source of carbohydrates. Derivative product of corn has beneficial for health such as maize starch, maize flour, and maize oil [1]. According to ASEAN Economic Community (AEC), corn is one of strategic commodities in ASEAN that become special attention besides rice, soybeans and cassava and Indonesia is one of the largest producer of corn in ASEAN [2]. The availability of corn in large scale would generate large waste of corn such as corn cob and it becomes a problem for the environment if it is not handled properly.

Corn cob is a lignocellulosic biomass sustainable in a bioethanol production because consist of lignocellulosic material. The main component of lignocellulosic in corn are 38.8% of cellulose, 44.4% of hemicellulose, and 11.9% of lignin [3]. Based on the content, corn cob has great potential to be hydrolyzed to glucose, is a monosaccharide that can be used as a raw material for processed into various derivative products such as chemicals, pharmaceuticals, glucose, and bioethanol [4]. The large lignin content in corn cob causes small glucose yields in the hydrolysis process, so pretreatment is needed to destruct the lignin structure to improve the efficiency of the hydrolysis process [5].

In addition, by utilizing lignocellulose biomass as raw material for bioethanol, food availability will remain stable. Another advantage of using this material are its abundant presence, environmentally friendly which does not produce greenhouse gas emissions such as petroleum fuels, and sustainable energy [6].

This research was carried out at chemical pretreatment using ionic liquid which has an advantage of not having an inhibitory effect on the enzymatic process [7]. Ionic liquids are environmentally friendly solvents which are stable in temperature, not volatile and can be reused (recycle) [8]. The type of ionic liquid used in this study is choline acetate (ChOAc) which has no inhibitory properties against enzymes and microorganisms compared to 1-ethyl-3-methylimidazolium acetate (Emi-mOAc) [9]. In addition to the ionic liquid pretreatment in this study also used alkaline pretreatment (NaOH). Treatment by adding NaOH can reduce lignin and can increase the accessibility of enzymes in cellulose [10] [11]. Therefore in this study the lignin removal process on corn cobs was carried out using the ionic liquids method using choline acetate (ChOAc) in combination with alkaline NaOH. This study aims to study the pretreatment results using ionic liquid, alkali of NaOH, and a combination of ionic liquid and alkali of NaOH to lignin decreasing and the level of lignocellulose crystallinity.

2. Material and method

2.1. Material

Material used in this study was corn cob, collected from agricultural land located in Bantul and Sleman, Yogyakarta, Indonesia. Corn cob was cutted into small size (1 cm^2) then dried under the sunlight for 2 days. And then it was crushed to become powder with 80 mesh of size and dried in the oven with 105°C of temperature to remove the water content bound in corn cobs powder. Ionic liquid used in this research was choline acetate (ChoAc) and for alkali method used NaOH.

2.2. Method

2.2.1. Ionic liquid method. Ten grams of corn cob powder was dissolved in 500 ml water and added with ChoAc in ratio 0,3 of IL : biomassa. The suspension was heated at 90°C, 120°C, and 150°C for 20 hours then filtered in centrifuge to separate the solid and liquid. The solid was washed by water and then separated from the liquid again. The water content was removed from sample using oven at 105°C until has constant weight.

2.2.2. Successive alkali method. Corn cob powder that has been pretreatmented by choline acetate was mixed with NaOH solution in ratio of solid : liquid = 1 : 20 (g/ml). Variation of NaOH concentration in this experiment were 1%, 3%, and 5%. The sample was heated at 90°C for 2 hours then filtered by a centrifuge to separate between solid and liquid. Solid was washed by water and separated again from the liquid and then ovened at 105°C of temperature.

2.2.3. Analysis method. In this study used two methods of pretreatment to remove lignin from corn cob powder were ionic liquid and successive alkali method. The ionic liquid used in this study was choline acetate (ChoAc) and the alkali was NaOH. Component of cellulose, hemicellulose and lignin were determined by chesson method [12] and characteristic of those were analyzed by Fourier-transform infrared spectroscopy (FTIR) and X-ray powder diffraction (XRD) method. Index of cristalinity was defined using Eq. (1) :

$$CrI = \frac{(I_{200} - I_{am})}{I_{200}} x100\% \quad (1)$$

CrI = Crystallinity Index

I_{200} = Crystal intensity at 2θ around 22.5°

I_{am} = Amorphous Intensity at 2θ around 18°

3. Results And discussions

3.1. Corn cob composition

Composition of untreated corn cob powder is presented in Table 1. As known in Table 1, it consisted of 35.82% of cellulose, 37.48% of hemicellulose, and 15.59% of lignin. In this study was found higher lignin than reported by previous study that was needed to be treated [3].

Table 1. Chemical composition of untreated corn cob powder

Component	Cocentration (%)
Cellulose	35.82
Hemicellulose	37.48
Lignin	15.59

3.2. Effect of temperature on corn cob composition in ionic liquid choline acetate pretreatment

Composition of corn cob treated by choline acetate in variation of temperature 90°C, 120°C, and 150°C are shown in Table 2. As indicated, choline acetate removed lignin 24.57% , 30.08%, and 36.05% in different temperature process were 90°C, 120°C, and 150°C, respectively.

Table 2. Chemical composition of corn cob after ionic liquid choline acetate pretreatment

Pretretment condition		Concentration (%)			Lignin removal (%)
		Cellulose	Hemicellulose	Lignin	
IL choline acetate	90°C	37.00	36.60	11.76	24.57
	120°C	35.70	35.90	10.90	30.08
	150°C	33.60	34.00	9.97	36.05

As seen in Table 2, treatment at higher temperature can removed lignin in higher amount. On other hand from Table 2 known that temperature can degraded cellulose. This occurs in treatments that used temperatures of 120°C and 150°C. From this treatment the highest cellulose content was found in choline acetate treatment at 90°C of temperature that was 37.00%.

3.3. Effect of combination of ionic liquid choline acetate and NaOH alkali pretreatment

The highest cellulose content in choline acetate treatment was then subjected to NaOH treatment. The NaOH concentration used were 1%, 3%, and 5%. Comparation of removal lignin and cellulose content between combination of treatment (choline acetate and NaOH treatment) and NaOH treatment seen in Table 3.

Table 3. Chemical composition of pretreated corn cob powder

Pretretment condition		Concentration (%)			Lignin removal (%)	Cristallinity Index (CrI)
		Cellulose	Hemicellulose	Lignin		
IL choline acetate + NaOH	1%	37.00	36.50	11.10	28.80	35.85
	3%	43.29	35.53	9.52	33.29	29.85
	5%	53.41	32.35	6.75	56.68	31.46
NaOH	1%	36.20	38.10	15.30	1.86	31.76
	3%	37.40	37.60	14.60	6.35	23.08
	5%	38.00	36.80	14.40	7.63	33.87

IOP Conf. Series: Materials Science and Engineering **625** (2019) 012029 doi:10.1088/1757-899X/625/1/012029

As seen in Table 3, combination of choline acetate and NaOH treatment removed lignin in higher amount than NaOH treatment. High concentrations of NaOH also has a higher lignin removal effect either in combination of treatment or NaOH treatment. The higher lignin removal occured in process that used 5% of NaOH, either in combination of treatment or NaOH treatment was 56.68% and 7.63%, rspectively.

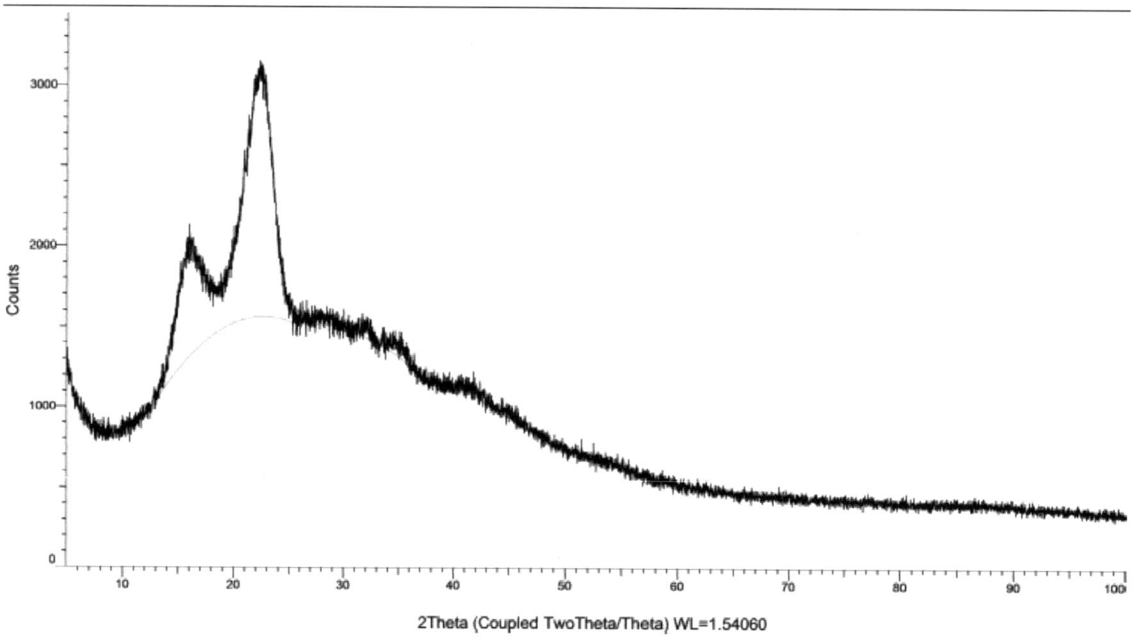

(a)

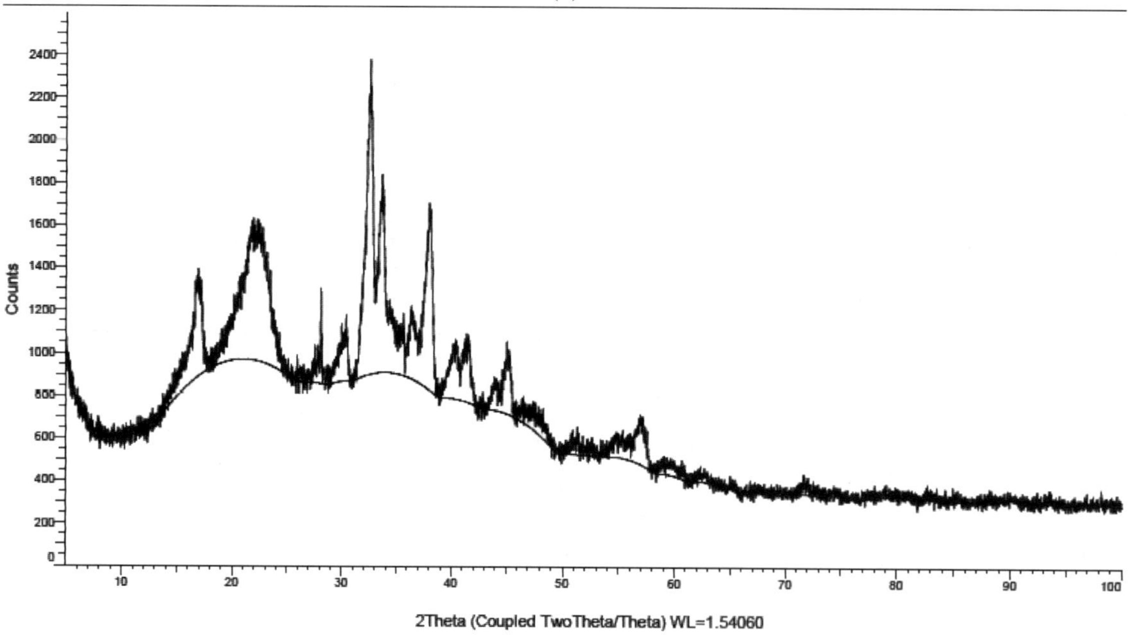

(b)

Figure 1. XRD spectrum of (a) combination of pretreatment (choline acetate and 1% of NaOH) (b) pretreatment 5 % of NaOH

Cristalinity index (CrI) and XRD spectrum can be seen in Table 3 and Figure 1. Table 3 shows that the highest CrI was obatined from process used concentration of NaOH 1% in combination pretreatment while 5% of NaOH in NaOH pretreatment. In combination of pretreatment the CrI was 35.85 while in NaOH pretreatment was 33.87. In the enzymatic hydrolysis process, the character of the cell wall and the level of accessibility of lignocellulose microfibrils become a factor that determines the hydrolysis rate. And enzyme accessibility is influenced by the crystallinity and size of the cellulose particles [5]. As indicated, an increase in crystallinity can occur due to removal of amorphous parts by alkali pretreatment. This is in accordance with previous research conducted by Williams 2011 [13].

4. Conclusions

In the current research many studies have been conducted to find environmentally friendly and sustainable energy as an alternative to fossil-based energy. To solve this problem, lignocellulose biomass is a concern of researchers because it is available in abundance. Large lignin content in lignocellulose biomass can affect the enzymatic hridrolization process so pretreatment is needed to eliminate it. From this study, combination pretreatment (ionic liquid choline acetate and NaOH) and NaOH can be considered as a method that can be used to obtain lignocellulose with a high cristalinity index and eliminate lignin. In the combination of pretreatment the highest lignin removal was found from process with 5% of NaOH as much as 56.68% but the cristalinity index was found from proces with 1% of NaOH as much as 35.85. While, In the NaOH pretreatment the highest lignin removal and cristalinity index was found from process with 5% of NaOH as much as 7.63% and 33.87, respectively.

Acknowledgments

The author would like to gratitude to Chemical Engineering Study Program of Universitas Islam Indonesia for the financial support and all of lecturers and staff so that this study can be completed properly.

References

[1] G. E. Orhun, (2013), International Journal of Food Science and Nutrition Engineering **3** (2), 13-16.
[2]. A. Sulaiman., (2015), https://www.indonesia-investments.com/news/news-columns/corn-production-consumption-in-indonesia-aiming-for-self-sufficiency/item5800? (accessed 29 January 2019)
[3] M. Pointner, P. Kuttner, T. Obrlik, A. Jager, and H. Kahr, (2014), Agronomy Research **12** (2), 391–396.
[4] Ó. J. Sánchez and C. A. Cardona, (2008), Bioresource Technology **99** (13), 5270–5295.
[5] S. Park, J. O. Baker, M. E. Himmel, P. A. Parilla, and D. K. Johnson, (2010), Biotechnology for Biofuels **3**, 10.
[6] Kumar, Parveen., Barret, Diane. Et al, (2009), American Chemical Society
[7] S. H. Mooda, A. H. Golfeshana, M. Tabatabaeib, G. S. Jouzanib, G. H. Najafi, M. Gholamib, and M. Ardjmanda, (2013), Renewable and Sustainable Energy Reviews **27**, 77–93.
[8] R.P. Swatloski, S.K. Spear, J.D. Rogers, (2002), J. Am. Chem. Soc. **124**, 4974-4975.
[9] K. Ninomiya, A. R. I. Utami, Y. Tsuge, et al (2018), Chemical Engineering Journal **334**, 657-663.
[10] H. Y. Li, X. Chen, Y. J. Li, et al (2018), Separation and purification technology **191**, 364-369.
[11] D. Chiaramonti, M. Prussi, S. Ferrero, L. Oriani, P. Ottonello, P. Torre, F. Cherchi, (2012), Biomass and Bioenergy **46**, 25–35.
[12] R. Datta, (1981), Biotechnology and Bioengineering **23** (9), 2167-2170.
[13] T. Williams, M. Hosur, M. Theodore, A. Netravali, V. Rangari, and S. Jeelani, (2011), International Journal of Polymer Science.

IOP Conf. Series: Materials Science and Engineering **625** (2019) 012030 doi:10.1088/1757-899X/625/1/012030

Assessment of climatic variation on bio cement mortar-based infrastructure

M Abdulkareem[1,2], F Ayeronfe[3], M Z A Majid[2], A R M Sam[2], J H J Kim[4]

[1]Construction Research Centre, Universiti Teknologi Malaysia, Malaysia
[2]School of Civil Engineering, Universiti Teknologi Malaysia, Malaysia
[3]Faculty of Engineering Technology, Universiti Tun Hussein Onn Malaysia, Malaysia
[4]School of Civil & Environmental Engineering, Yonsei University, Republic of Korea

Email: muyikareem@gmail.com

Abstract. Over the last few decades, effects of climate change have been widely studied by researchers. This change alters environmental conditions, which in turns influence the activities of bacteria. Therefore, this study assesses the effect of climatic variation on bio cement mortar-based infrastructure. This is achieved by evaluating the effects of temperature and relative humidity on the compressive strength of bio cement mortar. The bacteria used in the bio cement mortar are bacillus subtilis. 30 ml of bacillus subtiliss pore solution was added to the mix to create bio cement mortar. Test sample cubes of bio cement mortar were prepared and cured at different temperatures and different relative humidity. In addition, control sample cubes of cement mortar without bacteria were water cured. The 7- and 28-days compressive strengths were obtained. The results showed that the compressive strength of the bio cement mortar increased as both temperature and relative humidity increased. The 7- and 28-days compressive strengths increased by as much as 83.3% and 110% respectively when temperature increased, and by as much as 106.1% and 73.5% respectively when relative humidity was increased.

1. Introduction

Infrastructure are the most important and valued asset of any society. Yearly, governments across the world allocate a large chunk of their annual budget on maintenance [1-2]. Despite these budgetary allocations, deterioration in civil infrastructures remains a critical issue. The appearance of crack in civil concrete structures due to shrinkage, settlement and contraction causes to ingress of dangerous chemicals that may lead to severe damage [3].

Being the societal most valuable asset [4], researchers have come up with several concrete crack-sealing methods. These methods involve using epoxy systems [5], silicon-based polymer [5], chlorinated rubber [6] and acrylic resins [7]. However, the cost, compatibility and heat resistance has hindered the use of these crack-sealing methods. In view of this, self-healing concrete was developed to alleviate the problem of crack occurrence in concrete [8]. Self-healing method of concrete repair involves adding calcite-producing bacteria into the concrete [9]. The respiratory activities of the bacteria produce calcite which in turns fills in the cracks. In addition to filling-in cracks, the calcite produced also fill the micro pores in the concrete, thereby increasing strength, durability and density.

Several researchers have applied various bacteria to improve this property of concrete. Ramakrishnan [10] showed that *Bacillus pasteurii* increased the 28 days compressive strength of concrete by 18%. Similarly, Ghosh et al. [11] increased the 28th-day compressive strength of cement mortar by 25% when calcite precipitating bacteria was added to it. They stated that bacteria activity produced calcite within pores of the cement mortar. Achal et al. [12] improved the strength of cement mortar cubes by 17% when the cubes were treated with *Sporosarcina pasteurii*. The authors concluded that microbial calcite precipitated plugged the mortar pores, subsequently enhancing the compressive strength of the mortar cubes. Furthermore, Bundur et al. [13] increased the compressive strength of cement mortar by incorporating vegetative bacterial cells of *Sporosarcina pasteurii* into the cement mortar. A 23.5% increase in compressive strength was reported by Bensal et al. [14] by applying halophilic bacteria *Exiguobacterium mexicanum* isolated from sea water to concrete.

The incorporation of helpful bacteria has no doubt improved the self-healing and properties of concrete/cement mortar. It has also been shown by researchers that the environment influences the activities of bacteria [15]. Their calcite precipitation depends on factors like pH [16], nucleation site [17], dissolved inorganic carbon and calcium ions concentration [18]. The concentration of dissolved inorganic carbon depends on several environmental parameters such as temperature and the partial pressure of carbon dioxide [19]. Therefore, this study aims at investigating the effects of changing climatic conditions of bio cement mortar-based infrastructure. Cement mortar mixed with bacteria (bacillus substilis) are produced in cubes of 50 mm x 50 mm x 50 mm and cured at different temperature and in different relative humidity are examined. The 7 and 28 days compressive strengths of these bio cement mortar cubes are obtained and examined. In addition, control samples of cement mortar cubes are produced to serve as a baseline for comparison of the compressive strength.

2. Experimental materials

2.1. Fine aggregate and cement
Clean natural river sand obtained locally was used in this study. The sand was washed with water and sun-dried to remove the moisture content. The fineness modulus and density are 2.5 and 1750 Kg/m^3. The sieve analysis performed showed that the fine aggregate is well-graded. The cement used was Ordinary Portland Cement (OPC) available in the market. The OPC has BS 12 1996 specifications and a product of Cement Industries of Malaysia Berhad (CIMA).

2.2. Microorganism
Spores of a bacteria isolate obtained from the upper layer of soil in Kuala Lumpur, Malaysia were used in this study. The bacteria was cultured in a liquid medium that contained 10 g tryptone, 5.0 g yeast extract and 10.0 g NaCl in 950 ml H2O. This was autoclaved at 120 ^oC for 30ins. Upon inoculation on laminar flow, the culture was incubated at 37 ^oC on a shaker operating at 150 rpm for 48 hrs. The microbial self-healing agent consisted of the 48 hr-grown bacteria cells and the substrate which were then suspended in the distilled water used in mixing the cement mortar.

2.3. Test cube specimens
The fine aggregate and cement were thoroughly mixed and water was added to it. The mix ratio was 1:2 and water-cement ratio was 0.45. 30 ml of the bacteria spore solution was added to the mixing water. The cubes were casted in 50 x 50 x 50 mm moulds and compacted with a vibration machine. The cubes were demoulded after 24 hrs and placed in their different curing conditions. Table 1 shows the curing conditions (temperature and relative humidity) applied to the bio cement mortar, while the control specimen was water-cured. These curing conditions were applied to simulate climatic variations we have around the world. The cubes were removed after 7 and 28 days and tested for their compressive strength according to ASTM C39 by using a Universal Testing Machine (UTM).

Table 1. Curing condition

Temperature (^oC)	Relative humidity (%)
10	50
26	72
40	95

3. Results

3.1. Temperature effect on compressive strength

The results obtained from the compressive strength tests of bio cement mortar cubes that were cured under different temperatures are shown in Figure 1. These compressive strengths were taken on the 7th and 28th days. In addition, Figure 1 also includes the compressive strengths of water-cured cement mortar (WC). In Figure 1, it is observed that the compressive strengths of bio cement mortar cubes are higher than the cement mortar cubes (WC). The increased compressive strength is due to the activities of the calcite-producing bacteria in the bio cement mortar. These bacteria's respiration produces calcites that fill up micro spores in the cement mortar, thereby increasing strength. In Figure 1(a), the results show that increased temperature increased the compressive strength at 7 days. For example, the 7 day compressive strength at 10 °C is 34.6 N/mm^2, and this value increased to 45.2 N/mm^2 when the curing temperature increased to 40 °C. This represents about 30.6% increment. The 7 day compressive strength of the bio cement mortar increased to 45.2 N/mm^2 showing an increase by as much as 110% compared to the water-cured cement mortar (WC). Similarly, the 28 day compressive strength shown in Figure 1(b) indicates that the bio cement mortar cubes exhibited higher compressive strength than water-cured cement mortar (WC). The 28 day compressive strength of the bio cement mortar increased by as much as 83% compared to the strength of water-cured cement mortar (WC). The 28 day compressive strength of the bio cement mortar attained a value of 50.1 N/mm^2. Furthermore, Figure 1(b) shows that increased curing temperature increased 28 day compressive strength bio cement mortar. The compressive strength increased from 42.6 N/mm^2 at 10 °C to 50.1 N/mm^2 at 40 °C. These increments in compressive strength due to temperature increase are because the optimum temperature for culturing the bacteria (*Bacillus subtilis*) is 37 °C [20].

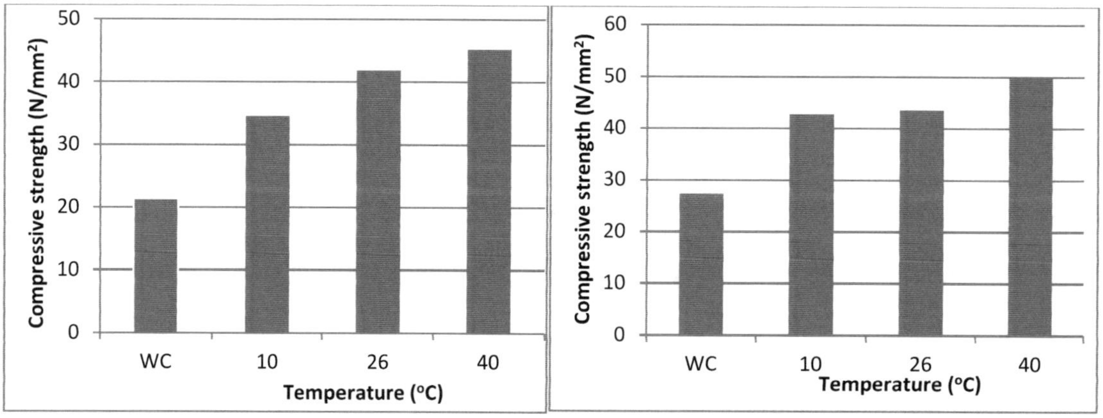

(a) (b)

Figure 1. Compressive strengths using varying curing temperature (a) 7 days; (b) 28 days

3.2. Relative humidity effect on compressive strength

This section reports the effect of relative humidity during curing on bacteria activities with respect to the compressive strength of bio cement mortar. The results explain the compressive strengths attained on the 7th and 28th days. In addition to the strength of bio cement mortar, the strengths attained by water-cured cement mortar (WC) at these times are also presented to serve as control. In Figure 2, the results of the strengths attained are shown graphically. Figure 2 shows that the compressive strength of the bio cement mortar cubes exceeded the control sample WC (water-cured cement mortar). For example, the strengths attained by WC were 21.4 N/mm^2 and 27.4 N/mm^2 on the 7th and 28th days respectively, while the minimum of the bio cement mortar were 40.1 N/mm^2 and 44.8 N/mm^2 respectively. From Figure 2(a), it is observed that increased relative humidity increased the early compressive strength. For instance, the early compressive strength increased from 40.1 N/mm^2 to 44.1

N/mm^2 when the relative humidity was increased from 50% to 95%. This indicates an increment of 10% at the early stage. Similarly, in Figure 2(b) that shows the 28 day results, the compressive strength increased as the relative humidity increased. At relative humidity of 50%, the compressive strength was 44.83 N/mm^2, and this value increased to 47.5 N/mm^2 at 95% relative humidity. This shows that relative humidity of bio cement mortar can be increased by as much as 6% when the suitable relative humidity is applied. The results in this section highlighting the effects of relative humidity on bio cement mortar show that relative humidity plays an important role in the compressive strength. The increased compressive strength experienced with increased humidity shows that the activities of the bacteria in the bio cement mortar was favored by this curing condition. This is similar to reports from several other studies that state that humidity positively affects microbial activities [21], thus more calcites were produced in the bio cement mortar. From the results presented in Figures 1 and 2, it can be seen that for the temperature and relative humidity considered in this study, both climatic factors have influence on the compressive strength of bio cement mortar at both stages. In addition, these two considered factors increased the strength at different manners.

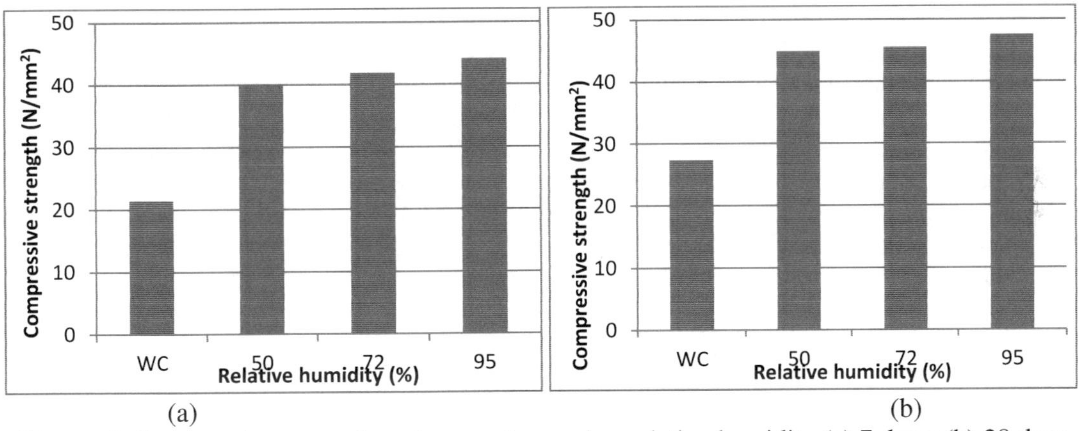

Figure 2. Compressive strengths using varying curing relative humidity (a) 7 days; (b) 28 days

4. Conclusion

This study evaluates the effects of climate variation on bio cement mortar-based infrastructures. This was achieved by assessing the 7 and 28 days compressive strength of the bio cement mortar. 30 ml of bacteria (*bacillus substlis*) spore solution added to cement mortar and cured in different climatic conditions. The climatic variation factors that were considered were temperature and relative humidity. The study have shown that irrespective of the curing condition, all the bio cement mortar cubes exhibited higher compressive strength than the water-cure cement mortar cubes (control cubes). This is due to the presence of bacteria that produced calcites to fill up microspores in the bio cement mortar. The compressive strength of the bio cement mortar increased by as much as 110% of the control cubes. Furthermore, it was observed that increased temperature and relative humidity increased both the 7- and 28-days compressive strength of the bio cement mortar cubes.

Acknowledgements

The authors would like to thank the Universiti Teknologi Malaysia (UTM) for their financial support through the Research University Grant (QJ130000.2409.04G00).

References

[1] Andersson A and Chapman L 2011 The impact of climate change on winter road maintenance and traffic accidents in West Midlands, UK. *Accident Ana & Prevention* **43** pp 284-289.

[2] Beuran M, Gachassin M and Raballand G 2015 Are There Myths on Road Impact and Transport in Sub-Saharan Africa? *Development Policy Review* **33**(5) pp 673-700.

[3] Samani A K and Attard M M 2012 A stress–strain model for uniaxial and confined concrete under compression. *Engineering Structures* **41** pp 335-349.
[4] Abdulkareem M, Bakhary N, Vafaei M and Noor M D 2016, Wavelet-based Damage Detection Technique via Operational Deflection Shape Decomposition. *Indian Journal of Science and Technology* **9**(48) pp 1-7.
[5] Pan X, Shi Z, Shi C, Ling T C, Li N 2017 A review on concrete surface treatment Part I: Types and mechanisms. *Construction and Building Materials* **132** pp 578-590.
[6] Schmidt M, Li L, and Spencer J 2001 Removal of chlorinated rubber coatings from concrete surfaces using an RF excited CO2 laser. *J of Materials Pro Tech* **114**(2) pp 139-144.
[7] Aguiar J B. and Júnior C 2013 Carbonation of surface protected concrete. *Construction and Building Materials* **49** pp 478-483.
[8] Jonkers H M, Thijssen A, Muyzer G, Copuroglu O 2010 Application of bacteria as self-healing agent for the development of sustainable concrete. *Ecological Engrg* **36**(2) pp 230-235.
[9] Zhang J, Liu Y, Feng T, Zhou M, Zhao L, and Li Z 2017 Immobilizing bacteria in expanded perlite for the crack self-healing in concrete. *Const & Build Mat* **148** pp 610-617.
[10] Ramachandran S K., Ramakrishnan V and Bang S S 2001 Remediation of concrete using micro-organisms. *ACI Materials Journal-American Concrete Institute* **98**(1) pp 3-9.
[11] Ghosh P, Mandal S, Chattopadhyay BD, Pal S 2005 Use of microorganism to improve the strength of cement mortar. *Cement and Concrete Research* **35**(10) pp 1980-1983.
[12] Achal, V, Mukherjee A, Basu P C and Reddy S 2009 Lactose mother liquor as an alternative nutrient source for microbial concrete production by Sporosarcina pasteurii. *Journal of industrial microbiology & biotechnology* **36**(3) pp 433-438.
[13] Bundur Z B, M J Kirisits and Ferron D 2015 Biomineralized cement-based materials: Impact of inoculating vegetative bacterial cells on hydration and strength. *Cement and Concrete Research* **67** pp 237-245.
[14] Bansal R, Dhami N K, Mukherjee A and Reddy M S 2016 Biocalcification by halophilic bacteria for remediation of concrete structures in marine environment. *Journal of industrial microbiology & biotechnology* **43**(11) pp 1497-1505.
[15] Ayeronfe F, Kassim A, Ishak N, Aripin A, Hung P and AbdulkareemM 2018 A Review on Microbial Degradation of Lignin. Advanced Science Letters **24**(6) pp 4407-4413.
[16] Achal V, Mukherjee A and Reddy S 2010 Microbial concrete: way to enhance the durability of building structures. *Journal of materials in civil engineering 23*(6) pp 730-734.
[17] Zhong W and Yao W 2008 Influence of damage degree on self-healing of concrete. *Construction and building materials* **22**(6) pp 1137-1142.
[18] Li V C and Yang E H 2007 Self healing in concrete materials, in Self healing materials. Springer. pp 161-193.
[19] Joshi S, Goyal S, Mukherjee A, Reddy MS 2017 Microbial healing of cracks in concrete: a review. *Journal of industrial microbiology & biotechnology* **44**(11) pp 1511-1525.
[20] Giri S, Harshiny M, Sen S, Sukumaran V and Park S C 2015 Production and characterization of a thermostable bioflocculant from Bacillus subtilis F9, isolated from wastewater sludge. *Ecotoxicology and environmental safety* **121** pp 45-50.
[21] Dannemiller K C, Weschler J and Peccia J 2017 Fungal and bacterial growth in floor dust at elevated relative humidity levels. *Indoor Air*, **27**(2) pp 354-363.

IOP Conf. Series: Materials Science and Engineering **625** (2019) 012031 doi:10.1088/1757-899X/625/1/012031

Characterization of coffee grounds as biosorbent for removal dyes from aqueous solutions

N. I. F. Mukti [1,*], and A. Hidayat[1]

[1]Chemical Engineering Department, Universitas Islam Indonesia
Jalan Kaliurang km 14,5 Yogyakarta 55584, Indonesia

[*]E-mail: nurindah_fm@uii.ac.id

Abstract. Low cost and effective biomass wastes may have ample applications in water decontamination procedures. Coffee waste is one of the most populer were used as low cost adsorbents for the removal of dyes from aqueous solutions. The aim of this research is to study the effect of washing the coffee residue for dyes adsorption. The characterization functional group of adsorbents was characterized with FTIR, and proximate analysis. The experimental data obtained in the present study indicate that coffee grounds are suitable for use as biosorbents in the removal of dyes.

1. Introduction

Indonesia is on of the largest coffee producing country in the world. In the last five years coffee production in Indonesia reached 650,000 tons/year, still below Brazil, Vietnam and Colombia [11, 22]. From the production of coffee, it will also produce solid waste of coffee grounds. In Yogyakarta itself many emerging coffee shops in recent years. So coffee grounds also increased in the region of Yogyakarta. So far, a small percentage of coffee grounds is used only as compost and animal feed, most of which are simply thrown away or burned away. The burning of these coffee grounds products contributes substantially to the production of carbon dioxide that can cause the greenhouse effect [23]. Therefore, it needs further handling in this coffee grounds for useful purposes and converting this waste into a new resource material.

Coffee grounds has a lignocellulosic composition of cellulose and hemicellulose is high enough. From some studies mentioned that coffee ground contain 51.5% cellulose and 40.55% hemicellulose [1]. From the composition of this coffee ground can be used as a raw material of adsorbent. Utilization of coffee grounds as a decontamination agent of chemical substances in wastewater has been developed. Coffee grounds pulp is widely used as an adsorbent because of its high adsorption capacity [13, 14]. One of them can be used as adsorbent to absorb the textile/batik dye from the wastewater waste of the textile/batik factory.

Adsorption is considered to be quite attractive in terms of its efficiency of dyes removal. Although the use of common materials (activated carbon, zeolite, clay) is still popular due to the high adsorption capacity, they are expensive, too. Thus there is a growing demand to find relatively efficient, low cost and easily available adsorbents for the adsorption of dyes. Some of the reported adsorbents include orange bagasse, palm fruit bunch, chitosan, etc [21, 7, 13]. Many literatures had studied about the possible application of coffee grounds as heavy metal adsorbents [13,10] and in particular as dye adsorbents [8,10, 15, 16]. Coffee grounds are generally called the solid wastes discarded from the extraction process of coffee beans.

In the current work, coffee grounds were collected from local cafe in Yogyakarta, Indonesia. The adsorption was performed in batch process and studying the effect of washing process of coffee grounds. The coffee grounds were also fully characterized.

2. Experiment

2.1. Materials
Coffee grounds were collected from local cafe (Hype Kulture and Kedai Rukun Lokal Cuisine & Brew, Yogyakarta, Indonesia). Remazol Yellow RG from Bratchem (Yogyakarta, Indonesia).

2.2. Preparation of biosorbent
The sample of coffee grounds were collected from local cafe (Hype Kulture and Kedai Rukun Lokal Cuisine & Brew, Yogyakarta, Indonesia) were washed with hot water to remove the brown color and residual organics and then dried at 80°C for approximately 8 h. Coffee grounds were also washed with metanol and n-hexane for comparing the effect of washing with the characteristic of adsorbent and the maximum uptake of the dyes. Fresh coffee grounds, Coffee grounds prepared by metanol and coffee grounds prepared by n-hexane are noted as CG, CGM, and CGH, respectively.

2.3. Characterization of biosorbent
Coffee grounds were characterized by lignocellulose content analysis using Chesson Method (Datta, 1981). Proximate analysis carried out by GB/T12496. Chemical characterization was carried out by FTIR spectroscopy in order to identify the functional groups at the surface of carbon materials. The FTIR measurement was carried out by FTIR Spectrophotometer (Shimadzu 8400S) using the KBr technique with wavenumber range of 500-4000 cm^{-1}.

2.4. Analysis of the sample
The dye concentration in aqueous solution was measured using absorption in a UV/Vis spectrophotometer at a characteristic wavelength. The amount of adsorption at the time, q_t (mg/g) was calculated from the following equation:

$$q_t = (C_o - C_t)\frac{V}{W} \quad (1)$$

Where C_o and C_t are the initial and final (equilibrium) concentrations of dye (mg/L), respectively, V the volume of dye solution (L) and W is the weight of the adsorbent (g).

3. Results and discussions

3.1. Biosorbent characterization
The proximate analysis for coffee grounds is shown in Table 1. From the results show the composition of coffee grounds is mostly carbohydrates of 62.14%.

Typical FTIR spectra obtained for CG, CGM, and CGH are shown in Figure 1. A comparative study of these spectra indicates that there was a slight difference in the results of FTIR analysis for CG, CGM and CGH. Two peaks at 2920 and 2850 cm^{-1} were identified in previous roasted Arabica and Robusta coffee samples (Kemsley et al., 1995) and also on Arabica green coffee samples (Craig et al., 2011, 2012), in relation to stretching bonds C–H is asymmetrical and symmetrical. Studies of FTIR analysis of caffeine on soft drinks have investigated peaks at 2882 and 2829 cm^{-1}, with the latter being due to the asymmetric stretching of C–H bonds of methyl (–CH_3) group in the caffeine molecule (Paradkar & Irudayaraj, 2002). Other FTIR studies also reported two peaks at 2927-2925 and 2855 cm^{-1}, being respectively attributed to asymmetric and symmetric C–H stretching in lipids (Cremer &

Kaletunç, 2003). Thus, the peaks at 2920 and 2850 cm^{-1} observed in the spectra presented for all samples (CG, CGM and CGH) in Figure 1 can be attributed to combination peaks to which both caffeine and lipids contribute. From Figure 1, it can be seen that at the wavelengths of 2920 and 2850 cm^{-1} for coffee treated with methanol and n-hexane deformed. It shows that the caffeine and lipid content in coffee grounds dissolves in solvents. In the wavelength area of 2850 cm^{-1}, a new peak appeared in the treatment using methanol and n-hexane which showed the addition of the $-CH_3$ group attached to the surface of the coffee grounds.

Table 1. Proximate Analysis of Coffee Grounds

Component	Composition, %
Carbohydrate	62.14
Protein	16.34
lipid	10.47
Ash	4.67
Water content	6.38

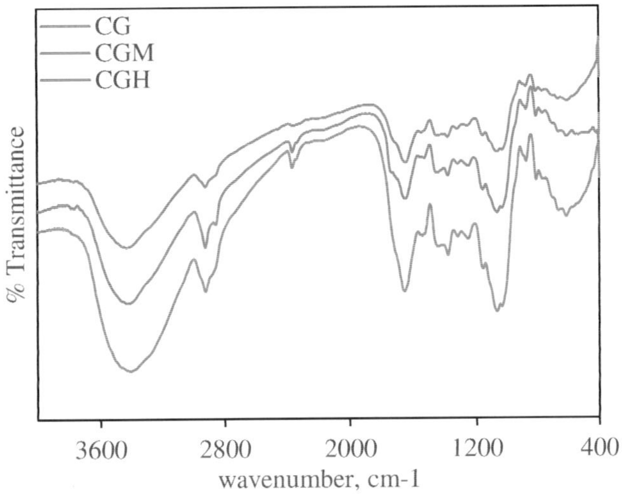

Figure 1. FTIR Analysis for CG, CGM and CGH

The peak at 1740 cm^{-1} was also investigated in the previous FTIR study in roasted coffee, related to carbonyl vibration (C=O) of the ester group in triglycerides (Kemsley et al., 1995) or aliphatic esters (Lyman et al ., 2003), show lipid bonds. The combination of absorption at 1740 cm^{-1} (stretching C=O) and at 2360-2260 cm^{-1} (stretching H–C=O) can be interpreted as a group of aldehydes (Miller et al., 2003), which are easily found in roasted coffee as a degradation of unsaturated fatty acids, such as linoleic acid, which is quite abundant in coffee lipid fraction (Oliveira et al., 2006). The results of this study, illustrate that the aldehyde group is also soluble in the solvents which are indicated by decreasing peaks in this area. The wavelength of 1650 cm^{-1} has been supported by Garrigues et al. (2000) because of the presence of carbonyl groups in their FTIR analysis of trichloromethane extracts from roasted coffee and then using quantitative analysis of caffeine procedures in samples of roasted coffee. In this study, it was seen that the peak of 1650 cm^{-1} was deformed. This shows that the solvent used can dissolve caffeine in coffee.

3.2. Biosorbent testing

The results of dyes (remazol yellow RG) adsorption on coffee grounds (CG, CGM and CGH) can be seen in Figure 2. From Figure 2, it can be seen that untreated coffee grounds (CG) is able to adsorb dyes better than coffee grounds treated (CGM and CGH). It is possible for the number of groups on the surface of the coffee grounds dissolves in the solvent so that the surface of the coffee can no longer adsorb the dyes.

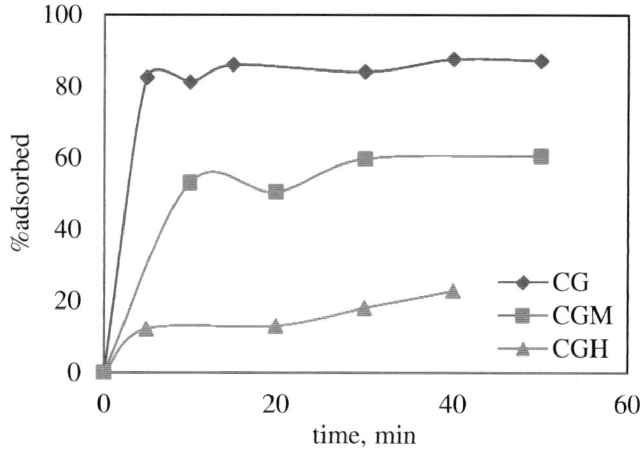

Figure 2. Adsorption of Dyes

4. Conclusions

Coffee grounds undergoes changes in functional groups on its surface after being treated using methanol and n-hexane. The functional group that showed caffeine, fat and aldehyde decreased in peak after treatment with methanol and n-hexane. The adsorption results showed that coffee without treatment was able to adsorb dyes better than coffee by treatment (CGM and CGH).

Acknowledgement

Funding for this research was provided by the Direktorat Penelitian dan Pengabdian Masyarakat, Universitas Islam Indonesia for financial support through the "Penelitian Pemula 2018" research grant and Department of Chemical Engineering, Faculty of Industrial Engineering for financial support for ICMENS 2019.

References

[1] Ballesteros, L.F., Teixeira, J.A., Mussatto, S.I., (2014), Chemical, Functional, and Structural Properties of Spent Coffee Grounds and Coffee Silverskin, Food Bioprocess Technol, Vol. 7, 3493–3503, DOI 10.1007/s11947-014-1349-z

[2] Craig, A. P., Franca, A. S., & Oliveira, L. S. (2011). Discrimination between immature and mature green coffees by attenuated total reflectance and diffuse reflectance Fourier transform infrared spectroscopy. Journal of Food Science, 76, C1162-C1168.

[3] Craig, A. P., Franca, A. S., & Oliveira, L. S. (2012). Evaluation of the potential of FTIR and chemometrics for separation between defective and non-defective coffees. Food Chemistry, 132, 1368-1374.

[4] Cremer, D. R., & Kaletunç, G. (2003). Fourier transform infrared microspectroscopic study of the chemical microstructure of corn and oat flour-based extrudates. Carbohydrate Polymers, 52, 53-65.

[5] Barret, E. P., Joyner, L. G., Halenda , P. P. (1951). The Determination of Pore Volume and Area Distributions in Porous Substances: I. Computations from Nitrogen Isotherms. vol. 73, 373-380.

[6] Datta, R., 1981, "Acidogenic Fermentation of Lignocellulose-acid Yield and Conversion of Components", Biotechnology and Bioengineering, Vol. 23 (9), pp. 2167-2170.

[7] Fiorentin L.D., Trigueros D.E.G., Modenes A.N., Espinoza-Quinones F.R., Pereira N.C., Barros S.T.D., Santos O.A.A., (2010), Biosorption of reactive blue 5G dye onto drying orange bagasse in batch system: kinetic and equilibrium modelling, Chem. Eng. J., 163, 68–77.

[8] Franca, A.S., Oliveira, L.S., Ferreira, M.E., (2009), Kinetics and Equilibrium Studies of Methylene Blue Adsorption by Spent Coffee Grounds, Desalination, Vol. 249, 267-272.

[9] Garrigues, J. M., Bouhsain, Z., Garrigues, S., & De La Guardia, M. (2000). Fourier transform infrared determination of caffeine in roasted coffee samples. Fresenius Journal of Analytical Chemistry, 366, 319-322.

[10] Jeguirium, M., Limousy, L., Labaki, M., (2017), Environmental applications of coffee processing by-products, Handbook of Coffee Processing By-Products, http://dx.doi.org/10.1016/B978-0-12-811290-8.00009-8

[11] Ministry of Agriculture RI, (2017), Production of Coffee based on Province in Indonesia, 2013-2017.

[12] Kemsley, E. K., Ruault, S., & Wilson, R. H. (1995). Discrimination between Coffea arabica and Coffea canephora variant robusta beans using infrared spectroscopy. Food Chemistry, 54, 321-326

[13] Kyzas G.Z., Kostoglou M., Vassiliou A.A., Lazaridis N.K., (2011), Treatment of real effluents from dyeing reactor: experimental and modeling approach by adsorption onto chitosan, Chem. Eng. J., 168, 577–585.

[14] Kyzas, G.Z., (2012a), Commercial Coffee Waste as Materials for Adsorption of heavy Metals from Aqueous Solutions, Materials, Vol 5, 1826-1840; doi:10.3390/ma5101826.

[15] Kyzas, G.Z., Lazaridis, N.K., Mitropoulos, A.Ch., (2012b), Removal Dyes from Aqueous Solutions with Untreated Coffee Residues as Potential Low-Cost Adsorbent: Equilibrium, Reuse and Thermodynamic Approach, Chemical Engineering Journal, Vol.189-190, 148–159.

[16] Lafi, R., Fradj, A.B., Hafiane A., Hameed, B.H., (2014), Coffee Waste as Potential Adsorbent for The Removal of Basic Dyes from Aqueous Solution, Korean J.Chem.Eng., Vol. 31(12), 2198-2206.

[17] Lyman, D. J., Benck, R., Dell, S., Merle, S., & Murray-Wijelath, J. (2003). FTIR-ATR analysis of brewed coffee: effect of roasting conditions. Journal of Agricultural and Food Chemistry, 51, 3268-3272.

[18] Miller, F. A., Mayo, D. W., & Hannah, R. W. (2003). Course notes on the interpretation of infrared and Raman spectra. Hoboken, NJ, USA: Wiley.

[19] Oliveira, L. S., Franca, A. S., Mendonça, J. C. F., & Barros-Junior, M. C. (2006). Proximate composition and fatty acids profile of green and roasted defective coffee beans. LWT e Food Science and Technology, 39, 235-239.

[20] Paradkar, M. M., & Irudayaraj, J. (2002). Rapid determination of caffeine content in soft drinks using FTIR-ATR spectroscopy. Food Chemistry, 78, 261-266.

[21] Nassa M.M., Magdy Y.H., (1997), Removal of different basic dyes from aqueous solutions by adsorption on palm-fruit bunch particles, Chem. Eng. J. 66, 223–226.

[22] Ramadhani, M. dan Putra, Y.M.P., (2017), Produksi Kopi Indonesia Didorong Jadi Peringkat Dua Dunia, yang diakses pada 3 April 2018 dari http://nasional.republika.co.id/berita/nasional/daerah/17/08/04/ou5ekj284-produksi-kopi-indonesia-didorong-jadi-peringkat-dua-dunia.

[23] Safarik, I., Horska, K., Svobodova, B., (2012), Magnetically modified spent coffee grounds for dyes removal, Eur Food Res Technol, Vol. 234, 345–350.